TO MY MOTHER,
Mayme McIntosh Higgins, whose remodeling
experience inspired me to write this book.

CONTENTS

Part 1: GROUNDWORK

1. GET CRYSTAL CLEAR ON WHAT YOU WANT
Discover the best way to conceptualize your vision to create a remodel you absolutely love.

2. INSPECT YOUR HOME
Learn the top five reasons why you must inspect your home before starting your remodel.

3. UNDERSTAND THE LIMITATIONS
Discover how permits, covenants, and zoning affect what you're allowed to do with your home.

4. ORGANIZE YOUR HUMAN CAPITAL
Learn about the types of pros you'll need on your team and how to screen them.

5. WORK WITH A DESIGN PRO
Discover the best way to work with a design professional to match your ideas to reality.

6. BUDGET REALISTICALLY
Discover painless ways to shave your budget.

7. FIND THE MONEY TREE
Learn about your finance and incentive options and how to strengthen your loan application.

8. CREATE A VISUAL ROAD MAP
Learn how construction documents capture your vision.

Remodel
SUCCESS

HOME REMODELING
DONE RIGHT, ON TIME, AND ON BUDGET

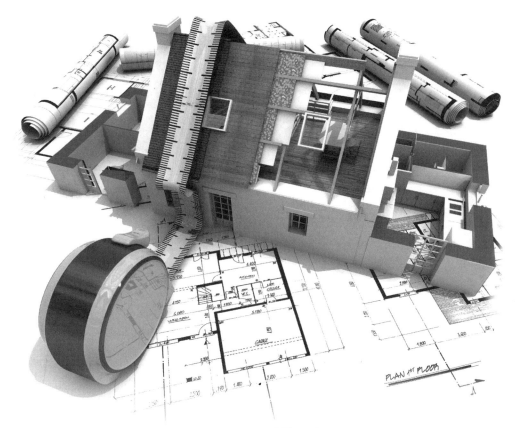

Monica D. Higgins

4880 Lower Valley Road • Atglen, PA 19310

Designed by Danielle D. Farmer
Cover design by John Cheek
Type set in Univers LT Std

Front Cover: Courtesy of Bigstock Images, 3D rendering of a house under construction with full technical details on top of blue prints, and a measuring tape © Franck Boston
Back cover: Author portrait by David Calicchio, Courtesy of Shutterstock, Interior Decorating © William Berry
Book Interior: Courtesy of Bigstock Images, Framed exterior wall going up on addition for existing home © Sue Smith, Computer Tablet Showing Finished Kitchen Sitting On House Plans With Pencil and Compass © Andy Dean Photography, People, fining documents and doing their administration, working on a laptop and filing invoices and bills in a ring binder, archiving them for tax returns © Corepics

ISBN: 978-0-7643-5405-2
Printed in the United States of America

Published by Schiffer Publishing, Ltd.
4880 Lower Valley Road
Atglen, PA 19310
Phone: (610) 593-1777; Fax: (610) 593-2002
E-mail: info@schifferbooks.com
Web: www.schifferbooks.com

For our complete selection of fine books on this and related subjects, please visit our website at www.schifferbooks.com. You may also write us for a free catalog.

Schiffer Publishing's titles are available at special discounts for bulk purchases for sales promotions or premiums. Special editions, including personalized covers, corporate imprints, and excerpts, can be created in large quantities for special needs. For more information, contact the publisher.

We are always looking for people to write books on new and related subjects. If you have an idea for a book, please contact us at proposals@schifferbooks.com.

PART 2: BUILD

PART 3: WRAP-UP

PREFACE

Remodeling is not suited for flying by the seat of your pants. It takes planning. Older homes may suffer from deferred maintenance, as well as hidden structural, safety, or code issues that can increase the complexity of a remodel. Many spec homes and older houses were not optimized for energy and water efficiency. Few homes use green building techniques such as passive solar for reduced heating and lighting costs, or sustainable/renewable/reclaimed materials.

If you're buying a home, resale value should also be on your mind, whether you're planning to live in the house as a primary residence or use it as an investment. In today's mobile society, few people live out their entire lives in the same house, making it important to weigh the cost of remodeling choices against their impact on resale pricing and appeal.

Maybe you fear the remodeling process and want a better outcome and fewer surprises in your next project. Those outcomes can be anything from staying on budget, to hiring a reliable contractor, to avoiding problems with permits and inspections, to selecting healthy and energy-efficient products and systems. Armed with solid information on the key remodeling success factors in this book, you'll gain the confidence to take control of your remodeling destiny.

Homeowner changes contribute to up to 30% of construction delays. Those changes can increase costs up to 50% or more. Industry experts estimate that 80% of these expensive homeowner changes are the result of poor initial planning, poor contractor selection, and ineffective communication. Once you understand how the remodeling process works from the contractor's perspective, you'll be in a much better position to plan your projects appropriately, select the right contractors, anticipate roadblocks, avoid costly changes, and incorporate cost-saving green elements.

Creating your dream home is possible, even if you know little or nothing about remodeling or have had a bad experience in the past. *Remodel Success* will give you a new level of understanding and confidence that will transform your relationship with contractors and suppliers, and put you back in control of your renovation project.

ACKNOWLEDGMENTS

I would like to express my heartfelt gratitude to my clients for entrusting me with the remodeling of their homes. I would also like to thank those who provided support as I completed this book:

Gail Martin of DreamSpinner Communications and Randy Peyser of Author One Stop for knowing this book would evolve at the right time with the right publisher and for helping me take a somewhat complex topic and make it simple, engaging, inspiring, and empowering.

Dylan Stewart of Dylan Stewart Design for translating the book's concept into a one-sheet that spoke volumes, and to Kathy Norris for your ongoing encouragement.

Cheryl Weber and Pete Schiffer of Schiffer Publishing for sharing my vision of helping homeowners who hire pros to achieve remodeling success.

Lily Alan and Grace Palmer for their assistance behind the scenes so I could keep the writing momentum going.

Emily Willoughby, Julia Teen, and Reva Kussmaul for reading every page of the manuscript with scrutiny and offering valuable constructive feedback.

Marcus Walker for your unconditional love.

INTRODUCTION
YOUR HOME, YOUR WAY

You've decided to remodel your home, but you're not sure where to start. The best place to begin is with a plan. Planning is to remodeling what location is to real estate—the essential component for success. Skipping the planning process is like building a structure without a foundation. Resist this temptation. Projects without conscientious pre-construction planning cost significantly more than projects with a well-thought-out plan. Construction experts know that homeowners are the biggest cause of cost overruns. In fact, more than 80% of expensive change requests come from homeowners who are not detailed enough in their remodel plan up front. As a result, homeowners cause up to 30% of construction delays and increase the cost of their own projects up to 50% or more.

When people hear "location, location, location," they think of real estate. When you think of remodeling, think "planning, planning, planning."

WORKING WITH THE REMODEL SUCCESS SYSTEM

This book is designed to share my system of thinking through a remodel plan. I'll share the same process I use with clients one-on-one. It's a proven system that I have refined through projects both large and small. I estimate that, on average, my clients save a minimum of 35% on their remodeling projects, compared to approaching their project without a solid plan. For example, if you do not "sanity-check" your $100,000 budget against the scope of work, you could end up spending $135,000.

On the other hand, if you've taken the time to carefully think through every aspect of your project, you would have spent no more than $115,000, with the $15,000 overrun resulting from fixing unforeseen conditions such as an undetected water leak causing mold or termite damage requiring reframing and other corrective work. Your pre-construction budget would have already accounted for this $15,000 because you would have made allowances for contingencies outside your control.

I'll show you how to put the same principles to use on your project. While I am unable to guarantee what you'll save in dollars and hours, I can guarantee increased confidence and peace of mind. For most homeowners, those benefits alone are priceless.

Older distressed homes have unique issues that can significantly add to the cost and complexity of a remodeling project. They may have major structural, safety, and code issues that have to be addressed before any upgrades can be considered. You may have to make up for years of deferred maintenance, damage sustained while a home stood empty, or even malicious damage caused by an angry former owner who stripped the property of fixtures, appliances, and copper plumbing. This book addresses those realities and shows you how to design a remodeling plan that gets the most from your budget.

This book is also designed to be interactive. If you just read it from cover to cover, you'll miss the point. I ask a lot of questions—the same questions I ask my clients. They're designed to walk you through the remodel visioning process step by step. Please take the time to answer each question thoroughly. Along the way, you'll also find "Monica's tips"—tips and tricks I've learned over the years—usually after I've done it the hard way. These tips may be shortcuts, alternative perspectives, or hints. Make the most of them!

If you're remodeling with a spouse, significant other, or business partner, each person should answer the questions separately. Schedule a quiet time to compare notes. You may be shocked by how different your vision and expectations are for the final product. That's great, because you have just uncovered the first roadblock to a successful project. The least expensive time to iron out your differences is right now, *before* you retain the services of architects, 3-D modeling experts, or a contractor. Working through these questions will help you and your partner come to consensus on the must-haves, the nice-to-haves, and what you can live without. Most importantly, you'll avoid the tug of war around trying to build one project from two mentally incompatible images, which can result in high cost overruns and fractured relationships.

 WHAT YOU WILL NEED

To get the most out of the *Remodel Success* system, you'll need some tools and a way of organizing them.

Virtual notebook

Capture research content and resources as you encounter them and organize them with Evernote, Google Drive, or other online tools. This will make it easier to share your notes or images with your project partners when you are not at your home computer. If you prefer, use a three-ring binder instead.

Digital folder

Create a folder for each chapter of the book. Upload your notes or completed worksheets into each folder as you go. You'll want to create a Resources folder where you will keep names and phone numbers of companies that have items you want and an Inspiration folder on Houzz, Pinterest, or other online photo galleries. If you prefer, insert divider tabs in a three-ring binder and create one tab for each chapter, then insert your notes or completed worksheets in each section as you go. You will also want to create tabs for a Resources and Ideas section for jotting down flashes of inspiration and storing images cut from magazines.

Smartphone or tablet

Keep this by your bed to capture those midnight eureka moments. You never know when inspiration is going to hit. Or use three-ring notebook paper.

Hole-punched pocket folders

These are great for storing pages torn out of magazines, pictures printed off the Internet, and other materials.

Hole-punched 8 × 11 envelopes

Use these to store small items like paint color cards, fabric swatches, and small photos.

3-D modeling program

Free, fast, fun, and intuitive programs such as SketchUp allow you to visualize your rooms or floor plans in 3-D and give you the ability to share models with your project team. Hole-punched graph paper can be used as a 2-D alternative for sketching your ideas.

Photo-sharing site or dedicated memory stick (optional)

If you're taking a lot of digital pictures of elements you like or downloading a lot of images, upload copies to a photo-sharing site or save them on a memory stick dedicated to your project. This will make it easy to share the images with your project partners when you don't have access to a computer or mobile device.

Digital voice recorder (optional)

Use this if you prefer to capture your thoughts quickly and transcribe them later.

Let's get started!

GROUND-WORK

1. GET CRYSTAL CLEAR ON WHAT YOU WANT

Discover the best way to conceptualize your vision to create a remodel you absolutely love.

I once worked with a homeowner who had a general idea of what she wanted for her whole-house remodel and addition. She had already started the process by hiring a general contractor to put it all together in a cohesive plan. However, at no time during the pre-construction phase did anyone ask this homeowner what she envisioned. So the plans she received were simply drafting and engineering drawings necessary to obtain building permits. They lacked practical considerations for comfort and functionality and did not address space planning, scale, lighting, or energy-efficient systems. Since this homeowner never had the opportunity to articulate her vision prior to construction, she ended up clarifying it after the remodel was in progress. Those changes delayed construction by many months and nearly doubled her remodeling budget. Unfortunately, I was brought in after the damage was already done.

To clearly articulate your vision, you'll need to fully understand your personal circumstances, what's essential, what's just nice-to-have, and the condition of your home. This is important, because unless your remodeling team holds the same mental picture that you have, your project won't turn out the way you hoped.

If you're remodeling with a partner, the two of you should answer the following questions independently. Then compare answers and come to consensus *before* presenting any ideas to other team members.

? BACKGROUND QUESTIONS

These questions are important because they get some housekeeping issues out of the way.

1. Is this your first home? Do you intend to use the house as a primary residence or an investment property? If the remodeled property will be your primary residence, how long do you intend to live in this home? If the remodeled property is already your primary residence, how long have you lived in this home?

2. Describe your family's characteristics, such as children and their ages, pets, hobbies, allergies, and/or disabilities as they relate to the home you have purchased.

3. Describe the features you love most about your home.

MONICA'S TIP:

Sometimes it is easier to express dislikes instead of likes. If that is true for you, start by writing down specifics about what you don't like, and move on from there to create a list of features that you desire.

4. Complete the sentence by writing down eight adjectives that describe you: I am . . .

5. Do you have any measurements, blueprints, surveys, architectural drawings, or related documents for your property? If so, make a list of these documents and where you can find them (this will be important later).

6. Do you have easements or any other type of design impediments, such as a septic tank?

7. Do you have a proposed time frame for the project? If so, do you plan to do it in phases, finish it all at once, do it yourself, or hire someone?

8. Do you have somewhere else to live during construction? Even a small remodeling project can be noisy, dusty, and potentially dangerous. Think about the impact on school and work, the schedules of infants or small children, and the wear and tear on everyone's nerves as well as safety issues. In the case of an extensive project, building codes may not give you the choice to tough it out.

9. What's your level of interest in the design process? Non-existent (you'd like to hand it off to an expert); minimal (provide input but that's all); average (be engaged throughout to discuss possibilities and new ideas); or extensive (you have very clear ideas and want to personally make most decisions).

10. Have you ever been involved in a renovation project? If so, list the projects.

11. If you've been involved in a renovation project before, was it successful? List why or why not for each project.

12. How important is it that your new home be energy/water efficient, incorporate green strategies such as passive solar design, or use sustainable/renewable/reclaimed building materials to reduce environmental impact?

EXPLORE YOUR VISION

A successful remodeling project requires a level of clarity that we do not necessarily use in our everyday verbal communication (maybe we'd avoid other problems if we did, but that's beyond the scope of this book). If you're remodeling your kitchen, for example, do you envision a kitchen for a gourmet cook with commercial-grade appliances, a built-in refrigerator, and marble countertops for making candy or pastries? Do you envision a cozy family gathering place for family dinners? How about an eclectic room for entertaining with everything that a hostess needs, plus great traffic flow for guests to circulate? All of these scenarios are "new" kitchens—but they're very different in style, function, and cost.

You *can* learn how to specify exactly what you want and how to ensure that everyone else on your remodel team correctly understands your vision. As you answer the following questions, quickly jot down what first comes to mind and leave room to make notes or add thoughts later. Again, if you're working with a partner, answer these questions separately, then compare notes and work out the differences.

First let's take stock of your existing space. Start in the morning and move through your typical routine, listing any flaws you uncover:

- Is there space to circulate?
- Are there any bottlenecks or inefficiencies?

Now mentally open the drawers and cabinets.
- What have you stored and where is it?
- Can you reach it easily?
- Must you move things to get to other items?
- Is there a place for items you use seasonally or only occasionally?

Think about privacy and light.
- Do you have any privacy issues with the traffic flow, windows, multiple users?
- Where is the light coming from throughout the day?
- How much natural light do the rooms get?
- What heating/cooling challenges do you have? Can you factor in passive solar strategies to make the house more comfortable while minimizing energy costs?
- If it's an older home, are there any charming features you want to preserve?
- If the new home was a spec house built with low-grade materials, are there fixtures, flooring, cabinetry, or countertops that you want to upgrade?
- Can the wiring and plumbing support upgrades such as modern appliances, sound, and wifi?
- What sorts of structural, safety, code, or deferred maintenance issues need to be addressed, such as foundation cracks, mold, and low water pressure? For example, foundation cracks can be caused by hydrostatic pressure, poor compaction, plumbing leaks, or tree roots and can require expensive excavation to fix. Prioritize these issues and budget for them first.

Beyond those must-haves, what do you envision your completed new home to be like?
- Which rooms are involved?
- What's the feel or mood—cozy, contemporary, retro, historic?
- How many people will use the rooms on a regular basis?
- What activities will these people engage in while in the rooms?
- Will these activities change by season?

Here are a few other issues to consider:

- **Children:**

  The ages and stages of children affect how space is used. Safety, ability to childproof, storage, cooking, stairs, and the location of exits is important with babies and small children. Teens and adult children want privacy and private gathering spaces.

- **Pets:**

  Think about the type, size, and needs of any pets you own or plan to acquire. How will they access and use the areas?

- **Older adults or returning grown children:**

  What impact will they have on the space?

- **Entertaining and overnight guests:**

  Do you host cocktail parties for twenty people or intimate dinners for eight? Do your guests stay overnight? Think about traffic flow, privacy, accessibility, and storage. Does your plan leave room for extra people?

- **Seasonal uses:**

  Do you need a clear path to the patio for summer barbecues? A mudroom for snowy boots? A big oven for Thanksgiving turkey? An island for parties and buffets?

- **Accessibility:**

  Could you use the space with your leg in a cast? Will there be guests or residents who need extra room to maneuver, extra lighting, or first-floor access due to a temporary or permanent physical limitation? If you plan to stay in the house as you age, could the remodel accommodate changes in your physical abilities?

- **Ambient noise and privacy:**

  How will activity in the remodeled area affect others in terms of noise or people walking through? Can you hear what's going on elsewhere if you need to, or can you (or others) hear too much? Can you close off the area for privacy, security, or convenience, if necessary?

- **Storage location:**
  Avoid having to store items at the opposite end of the house from where they are used. Consider foldaway and multi-use alternatives. Keep in mind the effects of temperature and humidity on stored items.

- **Windows, outside entry/exit, air flow:**
  Does the area have a view? Should the windows be operable or fixed? Does the area need its own entry/exit to the outside? How will airflow and temperature affect usability?

- **Proximity of area to related tasks:**
  If you multi-task, will you have to constantly run between rooms or floors or can you place related task areas near each other?

- **Feng shui:**
  If this is important to you, the envisioning stage is the time to build good *chi* into your plans.

- **Sustainability:**
  Now's the time to consider elements that contribute to healthy indoor air and resource conservation.

- **Housing values in your neighborhood:**
  Be careful not to remodel your new home out of the average price range for the neighborhood!

MONICA'S TIP:

Thinking about present and future needs and use is important because changes that are simple to incorporate in the planning phase are expensive to retrofit a few years later.

- **Are you expecting to buy new furniture or use existing pieces?**

  Will the pieces you want fit into the room and still permit traffic flow?

- **What level of finish materials do you desire?**

  For example, will you use ceramic tiles or asphalt shingles for the roof? Wood, stone, or laminate flooring?

- **For an older home,**

  does it make sense to add character and reduce environmental impact with reclaimed building materials?

Don't be discouraged if your mental walk-through uncovers issues. Celebrate the list of flaws you just uncovered before spending lots of money and review your expectations.

MONICA'S TIP:

Take time to spell out your vision in as much detail as possible (verbally and in writing) so your message is clear.

Realize that changes to one area of your home may have unintended consequences in other areas. For example, placing a craft room next to the kitchen may mean that glue and paint smells make dinner unappetizing. Adding a mudroom may create a bottleneck as all the outgoing morning traffic tries to squeeze through a small area. In short, think about how your remodel will affect every resident of your home, including pets, in these activity areas:

Cooking/eating
Bathing/toileting
Sleeping

Working/studying

Socializing

Entertaining/TV viewing

Indoor/outdoor play

Indoor/outdoor traffic and air flow

Storage capacity and accessibility

Housekeeping/laundry

Arrival/departure (daily and on weekends or special occasions)

Parking

A solid vision for your remodel is no accident or lucky break. It is formed by clearly identifying what you want and need from your home, based on how you actually live in it.

ADDING ON ISN'T ALWAYS THE ANSWER

Making improvements that increase resale value should be a consideration, whether you are a first-time homebuyer, existing homeowner, or investor. As a first-time or existing homeowner, the return on your investment is more likely to be viewed emotionally, whereas the return on your investment as an investor is more likely to be viewed financially. Any way you view the return on investment, smart space planning is key.

Adding square footage increases your construction costs, as well as energy costs and property taxes. So before adding square footage, think about how the existing floor plan could be reconfigured for a more efficient layout. For example, one of my clients wanted to add 1,000 square feet to a 1,900-square-foot three-bedroom, two-bath house. Since they were not planning on having a family, this seemed an odd request. After analyzing their space needs, we were able to reconfigure the floor plan and only add 350 square feet of livable space, as well as a garage with direct entry into the home. That saved 650 square feet and associated costs!

Ask yourself whether you really need more square footage or whether you can find the space by reconfiguring the existing floor plan.

MONICA'S TIP:

Don't be discouraged if it takes more than one try to get your space plan right. Remodeling is an iterative process. Now is the least expensive time to make changes!

CREATE AN IDEA FOLDER

As you work through the envisioning process, it helps to collect examples of the things you like and don't like. This way, you'll have tangible items to show your architect, interior designer, and contractor. Use the folders in your Remodel Success virtual notebook or pocket folders and the envelopes in your Remodel Success binder to store the items you collect.

Here are some things to go in your idea file:

- Photos from Houzz, Pinterest, or other online photo galleries or from books or magazines of houses, rooms, or features you like.

- Pictures of furnishings you plan to use in the room (your own and new pieces you want to buy). Include dimensions if possible.

- Decorating ideas such as curtains, wainscoting, built-in cabinets, faux finishes, and architectural features.

- Photos and product specification sheets for windows, doors, decking, appliances, or other purchased features.

- Color cards for paint or wood stains, fabric swatches, and wallpaper samples.

- Photos of light fixtures, built-ins, and large decorative items such as fountains, stained glass panels, or other artwork.

- Sketches of where you intend to hang artwork or place fragile or valuable items.

- Photos of storage solutions such as closet organizers, linen closets, and other essentials in the remodeled area.

- Floor plans from home plan websites, books and magazines, or hand-drawn from houses you've seen and liked.

- Printouts of web pages from sites such as HGTV or the DIY Network. When working with an older home, look for sites and articles specifically about retrofitting, modernizing, or remodeling a similar home. You'll want to print hard copies in case the URL changes.

- Sketch out where you intend to position appliances or electronics such as computers so you can factor in where outlets and special wiring need to go and assess the overall strain on your home's electrical capacity.

- If you will decorate specific areas for special occasions or holidays, take pictures of where you want to place a Christmas tree, hang garland, or make other decorating changes.

- If you want any special features such as wireless Internet or a media cabinet, save photos or reviews of applicable gadgets.

- Sketches (use the grid paper in your Remodel Success kit) to show room use, traffic patterns, furniture placement, general layout, outlet/electronic requirements, and seasonal or large decorations.

- Samples of flooring, wallpaper, tile, or other materials.

- Ideas you've gathered from touring open houses, showcase homes, historical estates, friends' houses, or past residences. Take pictures whenever possible.

- If you plan to incorporate feng shui elements, include your area grids and factor this into your sketches of room use and furniture placement. If there are architectural features you want to avoid, take pictures for your "don't" file.

- Ideas for working around a limitation you can't remove, such as a stairway or an interior support column.

DEVELOP A PRIORITY LIST

Homeowners rarely have a remodeling budget that allows them to do everything they want. That's why creating a priority list of must-have items and nice-to-have items is so important. Should the cost of materials go up, or should unexpected problems (such as finding a major structural problem) create extra costs, a priority list makes it easier to free up cash by letting go of features that really don't matter.

Start your must-have list with structural fixes, remedying safety issues, and bringing code violations into compliance. Also include anything that is absolutely necessary to make the home livable (working bathrooms, power supply to the kitchen, doors that lock, etc.). If the property is an investment rather than a primary residence, you still need to address these must-have issues, but your discretionary items will be based more on resale or rental value rather than personal use.

Now write down your must-have elements. Assign each element a number, where "one" is absolutely non-negotiable and the last element could be sacrificed if the budget requires it. When ranking items, consider how they will impact your quality of life and the lives of others who share the home. Do the same for your nice-to-have list. If you have a partner, do the rankings separately and then compare notes and work through any differences.

Things to consider as you rank your priorities:

- How important is this feature from an emotional standpoint?

- How important is it from a functional standpoint?

- How expensive would it be to add or change this later?

- How does this feature impact resale value?

- How much does this feature affect everyday quality of life or convenience?

- Does including this feature impact other costs? For example, a heavy light fixture may require additional bracing.

- Do I want this feature badly enough to give up something else?

- Will it bother me continually if I don't include this feature?

- Can I work around the function of the feature if I give it up? For example, if you leave out a warming drawer, can you use tabletop warmers instead?

- Will adding this feature negatively impact the home's resale value because it is unusual or overpriced? A great resource for sanity-checking your project is costvsvalue.com, which projects the payback on popular remodeling projects. It also contains details about your local real estate market, and lets you see how your costs compare to markets across the US.

- Can I live without recouping my investment if this item doesn't increase resale value?

- What's the life span of the feature? Appliances can be replaced more often than countertops. Perhaps you can trade off a feature with a temporary lifespan (faux finish, wallpaper, appliance) to keep a feature with a longer lifespan (custom wiring, built-in cabinetry).

- Will not having this feature really matter to me in a year?

- Will eliminating a green feature compromise any funding or incentives I am counting on in the financing?

- For investors, think about the cost of a feature that might result in a quicker sale versus the full cost of carrying an unsold house for an extended period of time.

- Is the feature a fad/trend or is it a fundamental?

Asking yourself these questions can be eye-opening. You may find that when you force yourself to choose, pricey items don't seem as important as features that will affect your everyday satisfaction. It's also a good way to get the compromising over with early, before a construction crew is waiting with the clock ticking.

Make a clean copy of your final prioritized lists so that there's no room for error. Make sure to get agreement from your partner before finalizing, if applicable.

Congratulations! By answering these questions, you are ahead of 99% of homeowners who undertake a remodeling project. As you move through the following chapters, come back and review your answers. It's okay to continue to make revisions and clarify the specifics. You'll also want to keep these answers in mind as you make decisions in later chapters. Use your priority list and idea file to remain true to your vision.

2. INSPECT YOUR HOME

*Learn the top five reasons why
you must inspect your home before
starting your remodel.*

Are you excited? Well, you should be, because now that your remodeling vision is crystal clear, you're ready to move forward! In this chapter, I'm going to show you how to avoid problems down the road by making sure you understand the condition of your existing home. I know you're in a hurry to get started, but trust me—rushing into a remodeling project is the ultimate example of penny wise and pound-foolish. A little extra time sizing up the situation right now can save you thousands of dollars, lots of headaches, and huge legal bills down the line.

Even if you've kept up with routine maintenance and have taken good care of your home, a home inspection prior to remodeling can uncover issues that might change your mind about whether to proceed. Now, I know what you're thinking. Aren't home inspections just for home buyers and sellers? Well, no. Home inspections are also for remodeling homeowners. Think of it as a crash course on the state of your home.

A home inspection is a general survey about the basic operation of your home and excludes systems or components that are not easily accessible or visible. Following are the top five issues a home inspection would uncover that may affect your remodeling project:

MOLD, RADON, AND ASBESTOS

That furry growth called mold comes in different colors and danger levels. Sometimes mold is easily visible—such as in a ceiling where there has been water damage. But all too often, mold is hidden behind drywall that appears to be intact, in the wood in your crawlspace, or in dozens of other places where water leaks or small drips can quickly add up to moisture damage.

Mold is a health hazard, one we're really only now beginning to understand. While individual sensitivity can vary, more and more research shows that certain types of mold can make you very sick. Some mold is dangerous enough that it can sicken someone with very little exposure. Other types of mold cause illness over a period of time, or make conditions like asthma worse.

Even if you haven't noticed health problems, it's not something you want to leave untreated. Mold gradually digests whatever it lives on, and if that's your house, that's a problem. Down the line when you want to sell your home, the presence of mold is almost guaranteed to either be a deal-breaker or the cause of an expensive remediation. You might as well bite the bullet now. Building over existing mold and trapping it in your new remodeling project is just asking for trouble.

If you're thinking of remodeling a fixer-upper you haven't purchased yet, finding mold should make you take a step back and think hard about finding a new house. Mold can be removed or remediated with abatement treatments that are both highly effective and reasonably priced. But rampant mold is expensive to treat and potentially impossible to get rid of completely, and a major mold removal project will be part of your home's permanent record. Make sure you know what you're getting into before purchasing a home to remodel that has mold problems, and make effective treatment (paid for by the seller) part of your purchase agreement.

Depending on the age of your home, nasty surprises may also be lurking beneath the old vinyl flooring or in the ceiling panels. While the mere presence of asbestos and lead is not necessarily hazardous, release of asbestos fibers and lead dust is. Asbestos is common in flooring and ceiling materials in homes built prior to the 1980s. Lead-based paint (used in homes built in the US before 1978) may have been covered by attempts to freshen up walls or woodwork. Renovation and repairs in a pre-1978 home require lead-safe-certified renovators to ensure that the dust does not spread.

Radon is a lung cancer–causing radioactive gas that you can't see or smell. It seeps through the ground, then into the air to make its way through cracks and other openings in your foundation and even your water supply. Have your home professionally tested (and mitigated, if necessary) by an EPA-qualified radon professional.

ELECTRICAL PROBLEMS

If you live in the home you're considering remodeling, take a look around. Are you using a lot of extension cords? Are outlets few and in questionable working order? Does plugging in extra lights or turning on major appliances cause a brownout? Do you blow a lot of fuses? Is your house more than fifteen years old?

If the answer to any of these questions is yes, you're likely to need an electrical upgrade, especially if your remodel calls for new major appliances or power-hungry additions like a hot tub.

When upgrading your electrical system, make sure the size of your electrical panel (the metal box from which electricity flows into your home from the utility company) will adequately accommodate your existing electrical needs with room for extra capacity should you need it. That means you need at least 200 amps.

BUILDING CODE

Making any significant change to your home may require bringing the entire structure up to code. If you bought a home from a reputable builder, you might think that code regulations don't affect you. However, codes change over time in accordance with public health and safety issues. Some of the recent changes include the number and type of smoke alarms and carbon monoxide (CO) detectors required, as well as the placement and number of outlets.

Sometimes the extent to which a remodel affects the rest of the house is up to the discretion of the inspector, and sometimes it is mandated by law. This means that remodeling one room of your home might necessitate upgrading your entire electrical panel, adding smoke detectors, or making other changes to bring your home up to current code. Those costs need to be known and planned for.

What about all of the do-it-yourself projects and handyman repairs done over the years? Unless you and the people you've hired were licensed and pulled the appropriate permits, you might find that those projects violate code, or even worse—present a fire hazard. Remember—if it costs too much to repair your fixer-upper, walk away.

PLUMBING

Low water pressure and drains that back up frequently are just two red flags that might indicate bigger plumbing issues. Sometimes the plumbing wasn't installed correctly in the first place, or repairs created the problem. In other cases, clogged pipes, leaking water mains, and water-hogging fixtures not only waste water (which costs money), but also result in water damage and the need for costly plumbing repairs. When you redo a kitchen, upgrade a bathroom, or install a tankless water heater, you may discover hidden plumbing problems that need to be addressed.

Pay attention to the location and route of sewer and water lines. Find out the distance from the main line. For example, when my home's main water line broke, it had to be repaired and reconnected to the city water line. While most homes on the block had a connection about twenty-five feet away, my connection was nearly four times as long. This was because my lot was originally the backyard of an adjacent lot that had been subdivided in the 1950s. This was a big headache and cost a lot of money to fix.

STRUCTURE

What you don't know about your home can hurt your health and safety, as well as your wallet. Hidden structural issues may not have caused problems yet, but they signal future trouble. A thorough inspection may help you uncover issues you wish you didn't have to deal with, but again, the bright side is that finding these problems early usually makes them easier and cheaper to resolve.

For example, foundation cracks could be an indicator of hydrostatic pressure, poor compaction, plumbing leaks, or tree roots, all of which you'll want to remedy. Look at the water flow from the roof. Where does the water go? Do you have problems with drainage near the foundation, standing water, or runoff during storms? Consider the impact of your desired addition. Will an addition trap water on the roof, making it more prone to leaks in that area? Can your foundation support the remodel you want to build? If you've been living with annoyances like slanted floors and doorways that aren't level, this might point to bigger problems related to poor soil quality or shoddy construction. Do you need to excavate to create your addition? If so, make certain you know what kind of soil or rock is under your lot. For example, expansive soils that absorb water and swell considerably can exert damaging force on a building. This type of soil can also shrink, removing support for the building load and causing damage.

If you're adding a room, can you make the new and existing roofs tie in nicely without replacing the entire roof? If your remodel requires opening up rooms by taking out walls, note the location of the load-bearing walls. If you plan on removing them, what is required to support the load-bearing beams and at what cost?

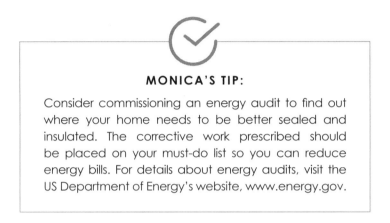

MONICA'S TIP:

Consider commissioning an energy audit to find out where your home needs to be better sealed and insulated. The corrective work prescribed should be placed on your must-do list so you can reduce energy bills. For details about energy audits, visit the US Department of Energy's website, www.energy.gov.

HIRING A HOME INSPECTOR

Ideally, you want to hire the most qualified home inspector you can find. Licensing or certification (which varies by state) is a must-have. If you can find an experienced inspector who is also a general contractor and who has estimating experience, you've hit the jackpot. An inspector with these credentials will be able to identify problems, estimate the scope of work needed to fix the issues, and give you an idea of whether your home's value justifies the cost of repairs.

Be sure to ask what the inspection covers. You might need to have separate inspections done, such as checking for termites. Make sure your inspection covers everything from top to bottom, including the roof, plumbing, attic/eaves, electrical systems, crawl spaces/basement, water heaters, windows and doors, fireplaces, water heaters, bathrooms, kitchens, balconies, pools, decks/patios, HVAC systems, built-in appliances, and exterior finishes such as siding and stucco.

The inspection report should contain photos, particularly of areas not easily accessible, such as under your home, the attic, and the roof. Add the required repairs to your remodeling budget. Remember: you'll make the best remodeling decisions when you have full information.

3. UNDERSTAND THE LIMITATIONS

Discover how permits, covenants,
and zoning affect what you're allowed
to do with your home.

PERMITS

The safety of building occupants is the primary reason for having building codes, and permits are the way cities help ensure that construction complies with those codes. Almost any kind of home improvement requires a building permit. Demolishing an old wing or addition is also likely to require a permit. Check your local building ordinances for details about code requirements.

Once you know the permitting requirements, you may decide to adjust your plans so you won't need a permit. For example, the code might require a permit for a fence more than six feet high. Can you live with a shorter fence? Discuss the options with the municipality so you don't make adjustments, only to discover that a permit is still required.

Make sure your contractor pulls permits, if required. If a contractor asks you to pull your own permits, find a different contractor. Why? Because a contractor is better prepared than most homeowners to navigate the permitting process and respond to building department questions, which in turn saves you time, money, and stress.

MONICA'S TIP:

Always obtain the requisite permits, inspections and documentation so you can get full credit for the work done and full value for an addition.

ZONING

You may have heard that good fences make good neighbors. So does good zoning. Zoning restricts what can be built in a particular area. It may reserve an area for single-family homes, or commercial properties, or multi-family dwellings (apartments and condos), or mixed use (a combination of residential and commercial). Zoning is supposed to help avoid problems by keeping someone from opening a noisy nightclub in a residential neighborhood, or putting a house in the middle of prime downtown commercial real estate.

While zoning regulations protect you, they also constrain you. For example, working from home is usually fine in a neighborhood zoned for single-family homes. However, if you intend to hang out a shingle and have customers traipsing in and out of your house/business and parking along the street, you're likely to run into zoning problems because you are creating a nuisance for your neighbors.

Fighting city hall is time-consuming and expensive—and usually futile. While areas can be re-zoned, it's a long legal process subject to public comment and politics. Zoning is generally good for everyone, so plan your remodel to get an outcome you and your neighbors can enjoy.

COVENANTS

While zoning is imposed by municipal government and works at a macro level, covenants operate on a micro level. They are enacted by builders and ultimately adapted by the neighborhood homeowners' association (HOA). Although covenants are not imposed by the government, they are legally binding. Many homeowners have come to grief because they did not read and understand the neighborhood covenants before they bought their home.

A covenant is a condition placed on the purchaser of a property that restricts the style, size, or type of building you may erect. Covenants are usually in place to ensure conformity to a certain standard. They are usually overseen by a neighborhood committee comprised of residents who hear petitions for projects that require approval or variances. Here are some of the most common covenant restrictions:

- **Size**

  The McMansion syndrome drove many homeowners to build huge homes that filled the entire lot, leaving little room between houses. In some neighborhoods, covenants cap the permissible house size to maintain a reasonable distance between homes.

- **Building materials**

  The types of exterior materials used on a home may be restricted, and even the disposal of demolished materials. This is especially true if you are in a historic district.

- **Temporary structures**

  Some homeowners prefer to live on-site in a mobile home or other temporary structure while their home is being remodeled to avoid the expense of a hotel. This may not be allowed because it would depreciate the value of surrounding homes. There may also be restrictions on the appearance of carports, detached garages, and garden sheds.

- **Time frame**

  Have you ever seen a home remodel that looked like an eyesore and took forever to complete? That's why covenants may restrict the time in which a remodel must be completed and the time in which landscaping must be completed after building/remodeling.

- **Vegetation**

  Covenants may prohibit the removal of vegetation, such as old-growth trees, that affects the neighborhood's appearance—something to consider if the new floor plan involves expanding your home's footprint.

• Fencing

Fencing restrictions may govern the construction type, height, placement, and areas fenced.

• Sale

Subdivisions may try to maintain a standard price for blocks of land, so sometimes the sale of a lot must first be placed with the developer to ensure that resales do not compete with new lots being released. Covenants may also forbid renting and leasing your home.

• Design review

If you are purchasing a home in a subdivision that has a design review board and plan to remodel it immediately, make sure there is a clause in your purchase contract stating that it is not subject to design approval. That way, if your dream home is not to the committee's liking, you won't be stuck with the house as-is. This only works if you are starting the renovation right after escrow closes and the city has approved your building plans.

MONICA'S TIP:

Don't become overwhelmed by these complexities. An experienced local design pro or contractor will have routinely handled this type of compliance and be able to figure out what works in your area.

4. ORGANIZE YOUR HUMAN CAPITAL

Learn about the types of pros you'll need on your team and how to screen them.

Major remodeling projects are not for the do-it-yourselfer. Reality home improvement TV shows often portray remodeling as quick and easy, with unrealistic project budgets. Television ads may try to convince you that you can do anything, but remember: they are run by companies that want to sell you building materials.

Even people who make their living in the construction trades are well-served hiring other experts with specialized knowledge. Working with a project team can help you avoid overwhelm because you have a group of experts to consult with when unexpected snags occur. With a team, you're less likely to be subject to one person's whims, and making progress is not entirely on your shoulders.

ARCHITECT OR DESIGNER?

What is the difference between a designer and an architect? An architect has earned a state-issued license to practice architecture by obtaining an architecture degree, completing an internship with a licensed architect, and passing rigorous examinations.

Once licensed, an architect is expected to uphold a professional standard of practice, support public health, safety, and welfare, and practice ethically. Only design professionals who pass the architecture registration exam can legally represent themselves as architects.

Designers are also professionals, but they are not typically licensed, though they may have earned a design degree. If the scope of your project affects your home's structural integrity or you are adding square footage, an architect will be required to stamp the drawings. Even if your project will ultimately require an architect's services, it might make sense to work with a designer in the early stages to develop an initial concept. Time is money and since a designer's hourly rates are typically less than an architect's, you won't burn through your design budget so quickly.

Either way, you would determine whether the design professional is a problem solver who listens and can also translate solutions into a space plan that: 1) aligns with your functional needs; 2) aligns with your aesthetics; 3) has successfully gotten numerous projects approved by the planning and building departments where the home is located; and 4) has drawn plans with a low rate of errors or omissions (less than 5%) discovered during construction. In other words, you want to hire the person who prepares the most accurate and understandable construction documents based on your budget. The person who can accomplish these things, regardless of whether he or she is an architect, is the right fit.

MONICA'S TIP:

The best way to predict whether the construction drawings will be complete, easy to interpret, and error-free is to ask the architect or designer for the names of contractors they've worked with. Then, ask those contractors if the project proceeded smoothly with few, if any, changes or disputes. You know what they say: You get what you pay for. Yet, the reverse is also true. What you don't pay for, you also get—in this case low-quality plans.

Be sure to vet the candidates thoroughly. Some specialize in certain types of work. For example, one may be great at creating a particular aesthetic while another may be great at space planning. A firm might specialize in kitchen remodeling, but not in building a new addition or altering the roofline for a dormer.

Some firms take a cookie-cutter approach and steer clients into a one-size-fits-all solution because it is what the firm is most comfortable doing. If that approach is not what you envision, choose another firm, regardless of how well-regarded the first group may be. This is your home, and the outcome should be what you desire.

Pay attention to how well the architect/designer listens to your program and concerns. The person should take copious notes, ask for clarification, drill down to details, and make on-the-spot recommendations or suggest appropriate alternatives to achieve your outcome and budget. You should feel that the two of you are on the same wavelength and that he or she "gets" what you are trying to achieve.

The best way to be happy with the design professional you hire is to be clear on exactly what you want from their services—and that requires doing your homework by following the steps outlined in this book. Having good chemistry also helps, so if you have to choose between two equally qualified people, pick the person with whom you get along with the best, since you're going to be spending a lot of time together.

Referrals from family, friends, neighbors, and co-workers are often the best way to find a good designer. At the end of this chapter, I am going to give you a list of questions to ask every professional you are considering using. You can also ask your friends why they were so happy with their experience, the kinds of questions they asked during the screening process, and the type of responses they received that led them to actually hire the pro. It's always a good idea to conduct an informational interview with architects or designers by phone to get a feel for where his/her specialties and interests lie. A person might have great skills, but lack interest in your particular kind of remodeling project.

STRUCTURAL ENGINEER

Depending on the project scope, it may be necessary to engage the services of a structural engineer. Your architect or designer will choose the engineer and work directly with him or her.

GENERAL CONTRACTOR

Your job needs someone in the general contractor role. If you don't hire a general contractor, you will have to do all those tasks yourself, which is a full-time job and requires attention to detail, a considerable amount of time spent on-site, flexibility to deal with last-minute issues (and there are plenty), and the personality to be hard-nosed when necessary to get tasks done on time. Most people either don't want the hassle or aren't suited for the job.

It has been my experience that finding good subcontractors is like herding cats. I prefer to find the good general contractors and let them do the herding. Why? Because when you have found a good general contractor, you have found a good pool of subcontractors. Your contractor will have built up relationships with tradespeople, know their reputation, and be aware of their strengths and weaknesses. Because those tradespeople also work in construction full-time, they will want to remain in good favor with the general contractor, because they know he/she can bring them other jobs. That works in your favor, especially when scheduling is tight or tradespeople are in high demand and must prioritize their time.

What does a general contractor do? The general contractor is responsible for assembling the team of trade specialists. This might include plumbers, electricians, HVAC professionals, tile installers, painters, cabinetry professionals, and other skilled workers. The contractor coordinates the schedule so that tradespeople come at the time their services are needed. Ideally, the tradespeople should not be on hand waiting around for their chance to work, nor should the project be held up waiting for them.

The general contractor puts together the master project plan, notes the critical path of what must be done, and notes all requirements for permits and inspections. He or she pulls the permits and coordinates the inspections, then handles any deficiencies noted by the inspectors and schedules the re-inspection. If tradespeople are unavailable, the contractor finds a replacement. When there is a snag in the schedule or an unexpected obstacle, the contractor figures out how to resolve it.

CONSTRUCTION MANAGER

If you fear being taken advantage of because you are not intimately familiar with the remodeling process, you may wish to have someone in your corner looking out for your interests. A construction manager oversees the project from pre-design through completion. This person knows what to look out for to make every penny count. Individuals become construction managers by obtaining a degree in construction science, construction management, building science, or civil engineering. Alternatively, they can obtain a certificate in the field regardless of their educational background. Whatever their credentials, you'll want to hire someone demonstrating a successful track record of getting their clients' projects done right, on time, and on budget.

What's the difference between a construction manager and a general contractor? General contractors and construction managers operate similarly. However, a general contractor will run your project from the perspective of building it, while a construction manager will run it from the perspective of being your project partner. This person will help you establish project goals that clearly articulate your vision, walk you through the pre-construction process, assemble a right-fit project team, and empower you to make good decisions by explaining your options. He or she will help you develop a realistic budget and reduce the occurrence of costly change orders. You'll get the most value from your construction manager by engaging him or her as soon as you begin serious planning. Make sure you choose someone who has significant experience with your project's scope of work.

? DO YOU NEED HELP...

- Clearly articulating what you *really* want?

- Formulating a realistic remodeling budget?

- Understanding both the design and construction processes?

- Selecting a designer or contractor who is right for your project?

- Understanding what is required to bring a damaged/distressed property up to livable standards?

- Clarifying the benefits and costs of incorporating resource-efficient elements?

If you answered yes to any of these questions, working with a construction manager to plan your remodel would be beneficial.

? WHAT'S YOUR REMODELING **PERSONALITY**

1. Before I begin a project, I always:

    a. Read everything I can about the process.

    b. Ask a few friends for their opinion or experience.

    c. Learn on the fly—experience is the best education.

2. Once a project is underway, I:

    a. Stick with it until it's done.

    b. Get frustrated when I hit complications.

    c. Have difficulty finishing it because I am easily distracted.

3. When it comes to communication skills, I:

    a. Have a proven track record of being able to give clear, consistent, and concise information to others.

    b. Often have to repeat myself or clarify—I don't seem to get through.

    c. End up losing my cool because they just don't get it.

4. When it comes to details, I:

    a. Am completely on top of it—I'm the kind of person they created spreadsheets and checklists for.

    b. Do a lot in my head—I have a good memory.

    c. Details aren't my thing.

5. When it comes to money, I:

    a. Can make decisions unemotionally and have experience tracking and sticking to a budget.

    b. Usually come out okay—I keep a mental running tally.

    c. I often argue with my partner about money. Budgets and accounts stress me out.

If your answers were mostly "a", you have a lock on all the details and probably have the tenacity to watch the nitty-gritty parts of your remodel. Be careful that you don't get lost in the minutia—you have a tendency to focus on the trees instead of the forest.

If your answers were mostly "b", managing your remodel will require self-discipline to deal with the small stuff, which bores you. You'll need to hone research skills, persistence, and write down details and costs to avoid problems. If your motivation is strong enough, you may be okay.

If your answers were mostly "c", you may want to rethink the decision to manage the remodel yourself. Details bore you and budgets frustrate you, and your "I'll know it when I see it" approach will cause your subcontractors to mutiny. You probably do many other things well, but a remodel is unlikely to be one of those things—hire help.

SCREENING YOUR TEAM

Protect your investment by hiring qualified team members. Make sure your project isn't their first such commission. You can be certain that, even if you have been offered an extremely attractive rate, there will be more than the usual delays and confusion due to their learning curve.

Ask to see proof of insurance, bonding, and workers' compensation coverage. Check with the Better Business Bureau, Angie's List, and other third-party review sites like Guild Quality or Yelp to see how satisfied other clients have been. Google the firm and see if they have been involved in disputes and lawsuits. Your local business magazine or newspaper probably runs a best-of ranking for professionals in various industries, although many good firms may not be large enough or have been in business long enough to be considered, and rankings alone are not a substitute for due diligence. Remember, qualifications are no guarantee of quality.

Here are the questions I ask candidates:

- Have you done similar projects? How many?

- What kind of projects do you prefer?

- How are your services or process different from your competitors?

- Do you have a particular philosophy or approach to your work and to the types of projects you accept?

- Can you provide references from clients who've done recent similiar projects? Even better—can you provide the addresses of previous jobs so I can drive by and see the work myself (for exterior jobs) and/or photographs to judge the quality for myself?

- Will I be working with a single person or a team? If a team, what are their individual specialties? How will I be billed for team input?

- What certifications and awards have you earned? Are you insured and bonded, and are you and your employees covered with workers' compensation?

This is just the tip of the iceberg. In chapter nine we'll deep dive into how to carefully assess your contractor candidates.

5. WORK WITH A DESIGN PRO

Discover the best way to work with a design professional to match your ideas to reality.

By now you have a clear idea of how you use your space. In this chapter, I'm going to show you how to create the environment you envision by working with a design professional who has space planning and 3-D rendering expertise. That way you can do a virtual walk-through well before construction starts.

The first step, though, is to produce scaled drawings of your home as it is now. This entails contacting your local department of building and safety or planning department to obtain a copy of building plans and other building records of your home (if you don't already have them) or having a draftsperson come to your home and measure each room and the lot. If the latter, start with having a floor plan drawn so you know what you are working with and can better see what you would like to alter.

MONICA'S TIP:

Take a stab at drawing your existing floor plan yourself. You'll need a long tape measure (at least fifty feet) or laser distance measurer and graph paper. Draw the perimeter shape of each room, measuring the length and width and the distance between the doors, windows, and other openings. Leave the other drawings to the pros.

While the draftsperson is drawing up the existing house, grab a camera. It's time to take photos of the current state of your home. Pretend you're a real estate agent who has to take pictures for potential buyers who have never seen the house. Get pictures of each room from every angle. Take photos of bathrooms and closets and from each end of the hallway. Take photos that show every wall, door, and window. Make sure you get photos that show the way appliances are situated as well as the location of sinks and plumbing fixtures. Don't forget the garage, attic, porch, and storage areas.

Print your photos or put them on a thumb drive for quick access when you meet with your architect and/or interior designer. Keep them in a safe place—you'll be referring to them throughout the planning process.

PLANNING YOUR SPACE

Space planning is more science than art because of the information gathering and analysis that must occur. These skills are honed over time, so work with a design pro who excels in space planning. You'll also want him or her to have 3-D rendering skills so you can see how your home is going to look. This makes it easier to evaluate different options as your project evolves and compare it to previous versions.

Once you have the floor plan drawings of your existing house, make copies that you can mark up without damaging the original. Make sure the floor plan indicates the location of lighting fixtures and outlets, specialty wiring receptacles for cable TV, land line phones or custom audio or media center equipment, and any other features such as whole-house vacuum connections. Now is the time to decide whether you want more outlets or outlets in different places, or for example, if you need to add additional ceiling light fixtures.

MONICA'S TIP:

Limit the design pro's scope of work to space planning. This will allow you to "test drive" whether he or she is a good fit for your project.

Grab your photos, the lists you made about how you use the space to be remodeled, the original floor plan, and your marked up copy. It's time to sit down with the design pro.

Your first meeting should be to discuss what you want from your remodel and walk him or her through your photos and floor plans. It's a good sign if the designer is happy about your preparation and is asking pointed questions, actively engaged in what you're sharing. The designer should speak with you about everything you uncovered when exploring your vision in chapter one. Mention it all—something you consider trivial could spark a great idea.

If you're fixing the house up to sell it, say so. The designer should have insights into what features are selling well and what current homebuyers look for, as well as which features might make it less attractive. Consider reaching out to your neighborhood realtor for further insight.

Have a heart-to-heart talk with the designer. There's no point in making a room look fabulous if it does not suit the way you live. I had clients who had previously remodeled their kitchen when center islands were starting to become the rage. They were advised to install an island so they'd have more workspace. Unfortunately, no one bothered to determine whether having a center island made sense and the best size and location. As a result, the person seated at the island would have to get up so the oven door could be opened fully. When it was time to remodel, that circulation problem needed to be fixed by repositioning the island so it ran parallel to the length of the kitchen. What a difference that change made to the function and aesthetics!

Based on the work you've done in previous chapters, the designer can develop up to three schematic floor plans (what your home would look like if you took the roof off). Once you've selected the space plan that will work best, it's time to visualize it in 3-D.

The color palette plays an important role in your remodeling vision. Think about what mood you want your home to evoke and select a palette that supports it. Then it's time to make decisions on product styles and materials. These include:

- Doors (interior and exterior) and windows (double-hung or casement, solid or side-lite entry door, etc.)

- Paint and wall finishes

- Wood trim and accents

- Flooring, countertops, and cabinets

- Plumbing fixtures for bathrooms and kitchen

- Major appliances

- Fireplace (gas or wood), mantel styles, hearth materials, fireplace doors

- Energy/conservation options, including Energy Star appliances, solar panels, tankless water heater, rain barrels, etc.

- Window coverings (curtains/shutters/blinds)

- Lighting (overhead fixtures, pole light, security lights, motion sensors, dusk/dawn sensors, etc.)

- Roofing materials and textures/colors, cornices, gutters, and soffit choices

- Siding options such as vinyl, beaded board, brick, stone, stucco, cedar shake

- Outdoor living areas and landscaping: pool, patio, deck, screened-in porch, walkways, retaining walls, fire pits/fireplaces/ovens, fences, trees and major plantings, drainage and irrigation.

MONICA'S TIP:

Keep things simple by starting with a "vanilla," black-and-white version of your 3-D space plan. This way you'll focus on form and function and won't get distracted by colors and materials.

An emerging trend in overcoming a design dilemma is the use of virtual reality (VR), particularly for those that have trouble visualizing 3-D renderings. With VR, you're able to immerse yourself in a space using a VR headset and move virtually through it. You're able to get a true sense of what it's like to stand there and whether any major changes are necessary.

Once you've decided which design scheme you'd like to pursue, your designer or architect will draw up the final plans. Request interior elevations so you can see things head-on. At this point, the designer should be able to provide you with an initial estimate. Realize that making changes, increasing the project scope, or substituting materials can dramatically affect the cost. Throughout the process, look for ways to maximize durability and get optimum performance from your building materials and techniques. This is another area where a good contractor can help you identify options and compare outcomes before you break ground.

Depending on the scope of work, it may be necessary to submit additional drawings for municipal approval:

Foundation plan
Roof plan
Site plan
Exterior elevations
Framing section

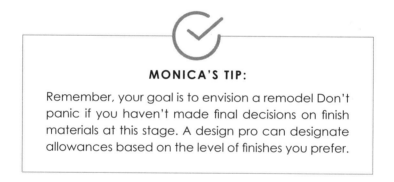

MONICA'S TIP:

Remember, your goal is to envision a remodel Don't panic if you haven't made final decisions on finish materials at this stage. A design pro can designate allowances based on the level of finishes you prefer.

6. BUDGET REALISTICALLY

*Discover painless ways to
shave your budget.*

A realistic budget puts the power in your pocket. You've heard the saying, "measure twice, cut once." That's a good metaphor for the budget, too. Once you've gone through the realization steps in chapter five, you should have a clear, detailed idea of what your ideal project will look like. Now it's time to crunch the numbers and see what your budget will support—before you start the bidding process.

MONICA'S TIP:

Remember, your goal is to envision a remodel you can afford to complete. So, be receptive to making adjustments that will help you reach the remodeling finish line.

FOUR REMODELING BUDGET BUSTERS

Unrealistic expectations can sink a project before it gets started. Unless you work in real estate or construction, you're unlikely to have a solid idea of your costs until you do your homework. Some people base their budgets on what previous remodeling projects have cost, what friends tell them they've paid for similar projects, or what they see on reality TV. The problem is, costs go up over time, so what a project cost five years ago has no bearing on what it costs now. Your friend's project might have used completely different materials, or been larger or smaller in scope, or been a do-it-yourself undertaking. Building and finish materials that are subtly promoted on home improvement reality TV shows are often donated to reduce production costs. In other words, you can't base your expectation of costs on anything other than an estimate for your specific project in the region where you live and at the time you want to do it.

1. Extras

Unless you have a very large bank account, you'll need to decide between nice-to-have and must-have features. If the estimate you receive after your in-depth design phase gives you sticker shock, refer to the priority list you created in chapter one.

2. Challenging site conditions

For example, if the property is on a hillside lot with limited access, chances are that large equipment will not be able to access certain areas of the property. As a result, your contractor will need to figure in manual labor costs since that phase will take longer when compared with using equipment.

3. Unforeseen expenses

Always factor in a contingency—a predetermined amount or percentage of the budget allocated for necessary, unpredictable expenses unrelated to expansion of the project's scope. Think of a contingency as protection against the unknown.

When budgeting for a contingency, you can either include it within the cost of each line item or it can be a single category of the overall construction budget. I prefer the latter approach because you can apply the contingency where needed. For example, some part of the project you thought might run over may pleasantly surprise you and come in at or under budget, while another item no one expected to cause a problem might create an issue that raises costs.

This is exactly what happened to one of my clients. The lower level of her home had water stains—evidence of a leak. However, the source of the leak was not determined until driveway excavation revealed a gap in the foundation. The existing foundation was built on piers with several feet between them.

Over the years, the dirt eroded, leaving this gap. When the exterior wall of the house cracked (most likely from settling or perhaps an earthquake) it created an avenue for the water to enter. My client was "covered" because a contingency was part of the budget.

4. Scope creep

This occurs when a homeowner hasn't clearly defined the project. It happens during the design phase and can even happen during construction, and many small changes can end up making a big cost difference. This happens innocently: You may think that since you're already redoing the bathroom, why not remodel the laundry room? Stop! Your scope is creeping. Be wary of phrases like "while we're at it," or "we might as well." They're warning signals that the scope is creeping. Once a project enters the building phase, new ideas become expensive change orders. Do your pocketbook a favor by nailing down your expectations in the design phase and sticking to it!

GETTING A HANDLE ON COSTS

Before you fall in love with the idea of remodeling your home, take a long, hard look at all of the costs involved. Here are the main costs to consider:

Professional design fees
Consultant fees
Permits/entitlements
Hard construction costs (materials and labor)
Landscaping
New furniture
New fixtures
Reserve against contingencies

When developing your remodeling budget, you not only need to consider the hard costs (materials and labor), but also soft costs such as architectural and engineering fees and permitting. Use the 20% to 29% rule: for every $10,000 in hard remodeling costs, budget $2,000 to $2,900 for soft remodeling costs.

Here are ballpark ranges for those soft costs, stated as a percentage of the total construction costs:

Architect/designer (10%–15%)

Consultants (6%–8%)

- Structural engineer
- Surveyor
- Soils specialist
- Geo-technical
- HVAC
- Title 24 Energy

Permits/entitlements (4%–6%)

- Planning / Zoning
- Building & Safety
- Public Works

Remember that at this stage you are still working with a conceptual budget provided by the designer that validates your proposed budget. In chapters ten and eleven we'll discuss the bid process, during which contractors provide a real number based on construction drawings.

The three key factors that affect budget are time, quality, and cost. I have yet to meet a homeowner who does not want quality work done fast and inexpensively. However, the reality is that you can only have two of these three criteria at one time. So pick the two you're most comfortable with. For example, if you pick quality and speed, just know that it's not going to be cheap. Why? Because if contractors do not have a large enough team to handle more than one project at a time, they'll need to forgo other projects to get yours done at warp speed. If you pick quality and low cost, prepare to be very patient, because your project is going to take a long time. By discounting

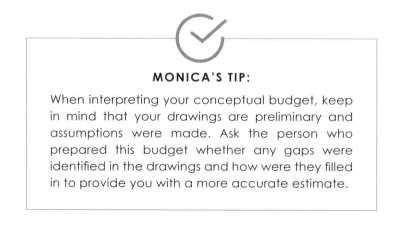

MONICA'S TIP:

When interpreting your conceptual budget, keep in mind that your drawings are preliminary and assumptions were made. Ask the person who prepared this budget whether any gaps were identified in the drawings and how were they filled in to provide you with a more accurate estimate.

price, contractors may not be as attentive to your project as they are for other projects paying market rates. You'll need to wait until their schedule clears. Finally, if you choose fast and cut-rate, you'll set yourself up to get exactly what you pay for—poor quality. There are plenty of people who claim to do framing, electrical, plumbing, or finish work well. However, my experience tells me that less than 2% do quality work.

You're going to make a lot of decisions that affect the final cost. By now, your design pro should have informed you of the cost impact of those decisions. Ideally, you want to track each decision and its impact during the design phase. When you're ready for a last pass, compare your original line-item budget to your projected budget based on your detailed plan.

If your design pro doesn't have a good grasp of construction costs, he'll need to bring in an estimator at this point to help ensure the proposed remodel fits the budget. I've heard too many horror stories of homeowners finding out during the bidding process that they can't afford what was drawn up for them. Getting an estimate during the design phase gives you an opportunity to tweak your plan early if the cost of your vision exceeds your budget.

A discrepancy between the estimated cost and the numbers you've been using thus far is a caution flag to stop and look more closely at your budget. If the value of any line item exceeds the value listed on the original budget by more than 10%, investigate the reason why. Did you change the scope of work or select a more expensive finish material? Have the detailed plans brought to light complications not previously recognized? Discuss ways to value engineer the project to determine lower-cost alternatives (adjust the scope of work or finish materials).

NINE WAYS TO SHAVE YOUR BUDGET

Here are nine great ways to reduce your project costs while keeping quality high.

1. Consider maximizing your tax write-offs with deconstruction instead of traditional tear-it-out demolition. Deconstruction is environmentally sound demolition in which used building materials are carefully salvaged from your home and donated. You'll need to hire a deconstruction contractor and an IRS-qualified used building materials appraiser, as well as determine which 501(c)(3) nonprofit organization will receive the donated materials.

Here's how it works:

a. Determine which materials you intend to donate.

b. Screen and interview an appraiser qualified to value salvaged building materials.

c. Verify the appraiser is independent of the donee organization. If not, your donation would be voided by the IRS.

d. Obtain a preliminary donation value from the appraiser.

e. Determine if the tax benefits of deconstruction plus avoiding the cost of traditional demolition pencils out. If so, proceed.

f. Pick a charity that focuses on salvaging donated materials. This will help you get the best value for your donation. Use Form 990 Return of Organization Exempt From Income Tax from the nonprofit or Form 1023 Application for Recognition of Exemption under Section 501(c)(3) of the Internal Revenue Code from the IRS.

g. Coordinate directly with the nonprofit to determine exactly what will be donated just in case there's something they don't accept.

h. Ask the nonprofit for deconstruction contractor referrals.

i. Interview and screen several licensed contractors with deconstruction experience. You'll want to work with one that will maximize the quantity of salvaged materials and minimize damage during removal. Don't forget to verify licensing, insurance, and bonding, and check references.

j. Select a deconstruction contractor you feel comfortable working with.

k. Complete IRS form 8283 for Noncash Charitable Contributions with your professional tax preparer.

The salvaged materials you donate will either be: shipped to a salvaged material retail warehouse like ReStore, run by Habitat For Humanity, for sale and distribution to the general public; broken into component parts that can be remade into cabinets, furniture, or flooring; and/or resurfaced by lumber mills for reuse in homes and commercial buildings.

2. Consider using recycled or refurbished materials if you can get the look you want cheaper than by using new or look-alike components. You'd be surprised what you can find at architectural salvage yards.

MONICA'S TIP:

The value of used building material donations can often pay for the costs of deconstruction. If your trash can become someone else's treasure, do well by doing good and pay it forward.

3. Stay within the existing footprint of your home. Bigger isn't always better. If you can achieve your dream home by making the existing floor plan more efficient instead of expanding, you'll avoid the costs associated with laying a new foundation for the addition.

4. Splurge only in areas used daily, such as kitchens and high-traffic bathrooms, that will give you the biggest bang for your buck. Go for look-alikes to brand name items in other areas where quality doesn't matter as much—for example, a powder room in the guest bedroom.

5. Keep the plumbing where it is. One little move could involve more work, costing more than it's worth. By focusing on cosmetic upgrades, you'll be able to reap the savings immediately.

6. Maximize natural lighting without windows. Tubular skylights are a compact, cost-effective, energy-efficient alternative to traditional skylights.

7. Ask if any of the bells and whistles you've added (textured glass panels, lighting or plumbing fixtures, decorative finishes) have more reasonably priced alternatives. It can't hurt to look, and you might find something you like just as well for a fraction of the price.

8. Look into federal, state, and local utility rebates and energy financing options for the use of systems and appliances that reduce energy and water use. This can sometimes save thousands of dollars.

9. Consider doing reasonably sized painting and clean-up jobs yourself (realize you will also be removing the contractor's accountability for the punch list). Consider this as a last resort. While it may seem great in theory, it may not be practical.

Also, consider hidden ways to shave off costs. For example, maybe the shelves inside your hardwood cabinets don't have to be made of hardwood. Perhaps plain switch plates will work just as well as those fancy designer plates. Be prepared to make compromises.

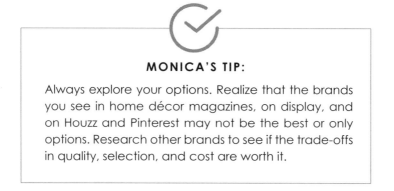

MONICA'S TIP:

Always explore your options. Realize that the brands you see in home décor magazines, on display, and on Houzz and Pinterest may not be the best or only options. Research other brands to see if the trade-offs in quality, selection, and cost are worth it.

7. FIND THE MONEY TREE

Learn about your finance and incentive options and how to strengthen your loan application.

There are many financing choices that can help make your remodel dreams a reality. If you're not planning to pay cash (and few people do), you'll need a loan. Banks look at your assets, liabilities, income, credit, and expenses to determine whether or not you're a good risk.

Before you apply for the loan, take these steps to strengthen your position:

• Check for credit report errors at least ninety days before applying for a loan. TransUnion, Equifax, and Experian offer free annual credit reports. Take advantage of this opportunity to confirm that every item on these reports is correct. If not, file a dispute with the appropriate credit bureau so they can investigate your claim. Send supporting documentation along with a letter. It will take approximately thirty days to hear back.

• Do not apply for new credit until after loan proceeds are disbursed. The excitement of remodeling lends itself to anticipating new furniture, but new credit inquiries may raise a red flag because the new credit increases your debt ratio.

- Hold off on closing paid-off credit accounts. Paying off credit is a wonderful thing, but closing those accounts may actually hurt your credit score. That's because while you're reducing the combined balance for all of your credit accounts, you're also reducing the combined limits. So, the percentage of credit you are using could increase, which may hurt your credit score.

- Offset the negative with the positive. Few people have perfect credit. Offset your less-than-perfect credit or other negatives with a loan-to-value less than the maximum allowed. If your debt-to-income ratio is a tad high, offset it with evidence of strong cash reserves of six months to a year or more.

- Gather income and asset documentation. Would you loan someone tens or hundreds of thousands of dollars without proof that they are able to pay you back? I didn't think so. While you'll need good credit and equity in your home, you'll also need to show financial stability as evidenced by pay stubs and tax returns. Missing or incomplete documents can hold up the process or lead to a rejection.

Strengthening your application for renovation financing positions you as an attractive applicant. It's not how good you or your home look in person, it's how good you and your home look on paper. Having a healthy credit score, paying down your debt, preparing your financials, and having sales comparables and a well-thought-out remolding process in place shows the lender that you're ready, are a low risk, and have the means to pay back the loan.

FINANCING OPTIONS

One way to finance a remodel is by borrowing against the equity in your home through a home equity loan or a home equity line of credit. What's the difference? When you take out a home equity loan, you're borrowing against the equity you've built up in your current house. The loan is often for a specific amount or a fixed percentage of the home's value. That amount is added to your existing first mortgage balance in the form of a second mortgage. You pay it back over time in equal installments over a fixed term, just like your existing monthly mortgage payment. This is an affordable option because it spreads the loan repayment over the term of your second mortgage.

Another option is a Home Equity Line Of Credit (some lenders call it a HELOC for short). This allows you to tap into a revolving credit line that works like a credit card. You only draw out what you need, when you need it, so you only have to pay back what you actually use. You'll make a minimum monthly payment (or more whenever you can)—usually interest-only for the first ten years. You'll need to have good credit for both types of financing and what the lender considers to be a reasonable loan-to-value ratio (LTV), which compares your home's value to the value of your loan. The higher the LTV, the higher the risk to the lender—hence, the higher the interest rate. Once your remodeling project is complete, a HELOC can provide additional flexibility by being available for other needs such as college tuition, at interest rates that are usually much better than those of credit cards. Or, you can keep the HELOC as a reserve in case of an emergency. However, if your home requires extensive work, this may not be a smart way to fund the project, particularly if you only intend to make the minimum payment.

Some lenders offer loans that cover the cost of purchasing/refinancing *and* renovating a home. This type of loan is based on the estimated after-renovation value of your home. With this all-in-one purchase/refinance and renovation loan, you typically make interest-only payments during construction. Once the work is completed, the loan converts to a standard mortgage.

There are a lot of advantages to a renovation loan or a purchase/refinance option. You can get started right away once closing is completed if your permits are ready. You're likely to get a higher loan amount since the amount isn't based on the current value of the home, but is calculated on the property's value after renovation. This kind of loan may also have a lower interest rate than credit cards or a second mortgage because it's technically a type of first mortgage. That helps keep your overall financing costs low, and by spreading the costs of your purchase and renovation over the loan period, you may be able to lower your monthly payments. Don't forget to take the tax deduction on your first mortgage interest, too!

A hard money loan may be ideal if you are a real estate investor who intends to sell the property upon completion. This type of loan is typically issued by private lending investors with rates that are usually higher than traditional loans because of the increased risk involved. While qualifying criteria for a hard money loan is similar to a traditional loan, most hard money lenders qualify a loan amount based on the value of the property, lending 60% to 70% of the property's after-renovation value.

For example, if the property is worth $100,000, the lender would lend $60,000–$70,000 for purchase and renovation. This leaves the hard money lender with sufficient equity in case you don't repay the loan and they have to foreclose on the property.

Think of a hard money loan as Plan B, when you are unable to obtain financing from Plan A—the types of financing mentioned earlier.

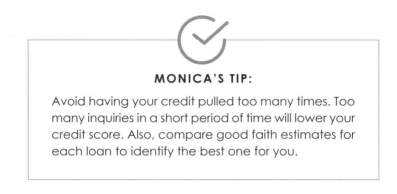

MONICA'S TIP:

Avoid having your credit pulled too many times. Too many inquiries in a short period of time will lower your credit score. Also, compare good faith estimates for each loan to identify the best one for you.

Think long and hard before you raid your retirement piggy bank to finance a major remodel. There's a lot to be said for using someone else's money. Even if you have sufficient cash in the bank to pay in full, you may want to consult your accountant and investment advisor to determine whether tax and asset appreciation repercussions make one financing choice better for your bottom line than another. Consider how long you plan to remain in the house as well as how near you are to retirement. All of those factors may impact your funding choice. Don't forget to check for special deals available for veterans, senior citizens, and credit union members if you qualify.

8. CREATE A VISUAL ROAD MAP

Learn how construction documents
capture your vision.

Now comes the challenge of tying your vision together in a series of written documents that describe its technical aspects. The information establishes the building parameters so your team knows exactly what you expect once your project is completed. The information shared in this chapter will walk you through these important documents and also help you confirm whether you can achieve what you want within your budget.

Construction documents include all building plans, specifications, and supporting documents such as structural, electrical, mechanical (HVAC), and landscape drawings. Specifications—the written requirements—are also included and cover the standards for products, materials, methods, and workmanship. They are prepared by your designer or architect, depending on the scope of work.

Construction documents are important because they translate your vision into a coherent written format easily understood by contractors. More important, they are incorporated into the contract between you and your contractor as the blueprint for building your vision. You'll use these documents to obtain reliable bids from contractors, and the contractor you ultimately hire will use them to obtain building permits. With all those uses, you can see why it's so important to make sure your construction documents do not contain errors, omissions, ambiguity, or conflicts, and that they reflect exactly what you want.

Most architects use CAD (computer aided design) to make drawings. Some also use BIM (building information modeling), a 3-D modeling database that shows the digital equivalent of building components and systems. One advantage of this technology is that whenever the model is modified, elements linked to that modification are updated automatically. BIM allows the walls to "talk to each other" so they can coordinate adjustments. This 360-degree approach provides critical insight before construction begins to help avoid costly oversights such as ductwork running into a beam.

Remember: if you can't see what you want on the drawings/specifications, it doesn't exist. Anything that isn't written down is open to interpretation (and misinterpretation), and without a paper trail, you've made an assumption instead of giving instructions. Assumptions are the first step toward big problems. Confirm, don't assume!

Depending on the scope of work, your construction documents will include some, if not all, of the following:

Site plan: How your site is or will be developed.

As-built drawings: The state of your home before remodeling starts.

Floor plans: Bird's-eye view of how the rooms are arranged.

Reflected ceiling/lighting plans: How lights and any other items that touch the ceiling, like access panels, are arranged.

Elevations: The front, sides, and rear of a building exterior.

Building sections and wall sections: A cut through a building to show its structural and construction elements.

Door, window, building details: The specifics about each of these items such as their measurements, type of material, etc.

Finish schedules, door schedules, window schedules: Lists comprised of manufacturer, color, type, model number, etc., for finish materials such as tile, flooring, door and drawer hardware, etc.

Interior elevations: The height from the floor for walls, windows, doors, and built-ins such as bookcases and fireplaces.

Millwork: Products such as baseboards, crown moldings, doors, window and door casings, trim, and banisters, that are typically produced by a wood mill. However, many of these products are made out of plastic or a composite material.

In addition, your construction documents will include:

GENERAL CONDITIONS: Boilerplate language about conditions applicable to most projects.

SUPPLEMENTARY CONDITIONS: Legal, physical, and/or climate conditions specific to your project that take precedence over general conditions.

SPECIAL CONDITIONS: Conditions specified by the homeowner, if any.

MONICA'S TIP:

When it comes to finish materials, always have a Plan B or even a Plan C in case it becomes necessary to substitute an item. For example, if at the time you're ready to place your order you find out that your first choice of an item has been discontinued or has a long lead time that would delay construction, you can quickly present a substitute.

Why do you need both drawings and descriptions? Properly prepared documents complement each other. Many construction professionals are visual thinkers, and that means drawings are the method of choice for showing the work to be done. The problem with drawings is that they are subject to interpretation. So while a picture is worth a thousand words, people do see things differently. Your construction documents provide detailed written descriptions of the work to be done. That way, if there's a question based on a drawing, the written document will make things clear.

When reviewing your construction documents, use the four C's to help ensure the project plans align with your vision.

C L E A R — C O N C I S E — C O M P L E T E — C O R R E C T

1. Are the drawings and documents **clear**? Is the information logical, and does it make sense?

2. Are the documents/drawings **concise**? In other words, are they focused yet comprehensive in scope?

3. Are drawings and documents **complete**? The items and elements covered in the drawings should be covered in the documents, and vice versa.

4. Are the documents and drawings **correct**? Do they provide sufficient detail or information to build what you've specified? Does data appearing in two or more places match?

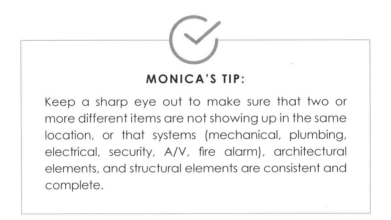

MONICA'S TIP:

Keep a sharp eye out to make sure that two or more different items are not showing up in the same location, or that systems (mechanical, plumbing, electrical, security, A/V, fire alarm), architectural elements, and structural elements are consistent and complete.

Some new details will inevitably arise during construction. However, you don't want these conditions to occur because of poorly developed construction documents. They should be the exception and not the rule. Clear, comprehensive construction documents will help contractors provide you with reliable bids without needing to ask for further clarification.

Preparing documents and drawings (and thoroughly reviewing them) may seem tedious when you're eager to get going on your project, but good documentation can make the difference between an on-budget dream remodel and a budget-busting nightmare!

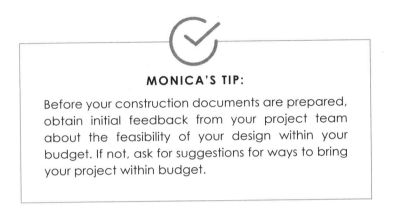

MONICA'S TIP:

Before your construction documents are prepared, obtain initial feedback from your project team about the feasibility of your design within your budget. If not, ask for suggestions for ways to bring your project within budget.

PART 2

BUILD

9. CHOOSE A CONTRACTOR

*Discover the most effective way to
"speed date" a contractor.*

While screening contractors takes time, it doesn't have to be overwhelming. As you go through the interview process you'll pick up on signals that will help you quickly determine whether or not a contractor is a good fit for your project.

I am a strong believer in asking behavioral interview questions because they are more probing than just asking for numbers and dates. Questions that involve why, what, and how require the contractor to respond with specific examples based on behavior and situations. As you interview, look for evidence of good business experience and sound management ability, expertise with your project type, excellent customer service, and clear communication. You'll also want to find out what kind of feedback previous or current clients can provide.

As you interview contractors, pay attention to tone as well as content. Is the contractor impatient with your questions, or worse, defensive at having his/her expertise challenged? If the contractor brushes you off, is condescending, or becomes irritated at being asked for specifics, it won't get better later. If you detect a negative attitude about your age, ethnicity, gender, or living arrangement, find another contractor. Bias is difficult to prove legally, but all too easy to recognize when you're on the receiving end. There are other fish in the sea. Find a better match.

If you have conservation-related priorities, now is a good time to ask about the contractor's experience with those aspects and also to discover if they are accredited or certified in green building programs such as Leadership in Energy and Environmental Design (LEED) or another respected independent third-party program.

General questions should include the following:

- How many years of industry experience do you have?

- Tell me about some of your favorite projects. What inspired you to become a general contractor?

- How have you kept current in your field?

- How many employees work for your company?

- How will you determine which subcontractors to use on this project, and what qualifications do you look for? How much experience have you had with these subcontractors?

- What types of insurance coverage do you carry? What are the dollar limits on each policy?

- Why do you want to work on this project?

- How do you establish trust with clients?

- How would your clients describe you?

- Do you love what you do? Why?

Experience questions hone in on the contractor's familiarity with your project type. Here are some of my favorite experience questions:

- Have you recently remodeled a home involving a similar scope of work? When and what was your most recent project? Can you provide examples of other projects you've done that are like my project?

- Will you provide a list of references (including names, addresses, and phone numbers) of clients for whom you've done projects like mine?

- How did your actual cost compare to the budgeted cost for projects similar to mine? Why the difference?

- Describe a challenging remodel situation. How did you meet the challenge?

- Tell me about a situation when you anticipated problems and developed preventive measures.

Services questions provide insight into the customer support, communication, and warranties offered. Ask the following questions:

- Do you warranty your work? If so, for what period of time? What is and is not covered?

- How will you determine a construction schedule? Based on what you've been told about this project, what do you feel is an appropriate schedule (weeks or months)?

- Do you provide a written timetable for when each phase of the work will start and be completed?

- How often do your projects adhere to your baseline schedule?

- Tell me about a situation in which you were unable to meet a deadline or goal. Why was the deadline or goal missed?

- How many projects does your company work on concurrently? How do you juggle multiple projects? This is important because many companies that do both indoor and outdoor remodeling trade off between projects based on weather. If you're in an area where good weather is limited and you have an indoor project, expect to lose the crew to roofing jobs or outdoor construction on every sunny day. This also sheds light on whether the contractor is in a position to take on a new project once permits are pulled and you're ready to start construction.

- Based on what you've been told about this project, what do you feel is an appropriate staffing level for your proposed schedule? How did you arrive at this number?

- How often do you provide schedule updates?

- How do you handle communication between your client and subcontractor team?

Talking about money may be uncomfortable, but it's essential. Here are some fee-related questions to get you started.

- How do you estimate your fees for a project and which services are included in those fees? What will the fee schedule be?

- How would you define a change order? How do you deal with change orders?

- What would be considered additional services and reimbursable expenses? How do you establish your fees for additional services and reimbursable expenses?

DUE DILIGENCE

The best general contractors have a stable of reliable and professional trade contractors with whom they work on a regular basis. Rather than focus on lowest price and risk dropping quality standards, these contractors have their best subcontractors visit your home for preliminary pricing feedback, then take your plans and specifications to them for closer review and negotiate a fixed price for their work. After negotiations, those numbers are included in the contractors' bid and the subcontractors are held to their original estimates. Ask if your candidate has a favorite list of subcontractors, and whether your job will be staffed from among them.

Far too often, we hear horror stories about a homeowner's experience with a contractor. Unfortunately, this has led to a misconception that the home improvement industry is rife with unscrupulous contractors. Don't let reports of a few bad apples keep you from moving forward. Instead, empower yourself by doing your due diligence.

One method for finding contractors is to ask owners of successful renovations in your area for their recommendations, then interview several of these contractors until you find three you would feel comfortable working with. The saying "good things come to those who wait" is true in this case, so be patient and don't settle for anyone with whom you don't feel totally comfortable.

When checking references, ask questions about how easy it was to communicate, how well the contractor got along with neighbors, and whether there was confusion over who was handling small details (like choosing and buying outlet covers). Don't forget to check on how well, overall, the contractor and crew behaved. This can provide valuable insight into the firm's general business principles.

Don't stop at speaking with references—visit them to see the quality of work first-hand. Take a look at recent projects as well as projects completed several years prior so you can see how they held up with normal wear and tear.

Verify with the state licensing board that the contractor is licensed for the work to be performed. Make sure the license will remain active throughout the construction phase and that the contractor has a clean record. Obtain copies of the contractor's insurance certificates to verify that the contractor carries current insurance in the appropriate levels to protect you from claims arising from job site injuries or property damage.

Check with your local Better Business Bureau or consumer protection agency to determine whether any complaints have been filed against the contractor, and the status of these complaints. More telling than a complaint is whether the contractor has a track record of not resolving complaints. Look for liens and judgments that suggest the contractor doesn't pay his vendors, crew, or subcontractors on time. Google can be your best friend when researching this kind of information.

Every homeowner wants a good job, at a great price, done quickly. The reality is that you cannot have all three, so most homeowners focus on price at the expense of build quality. Ask yourself which two of the three criteria are important and keep them in mind when interviewing contractors and obtaining bids.

Finally, avoid contractors with poorly written contracts that leave room for misinterpretation or those requesting more than the state-mandated minimum fee to start your project. These are signs of a poorly managed business. If you've been working through the chapters in this book, you should have very detailed documentation and drawings, so communicating what you expect with clarity and precision on your end shouldn't be a problem. Make certain the same is true on the contractor's end.

10. CONDUCT A PRE-BID MEETING

*Learn the best way to
get an accurate bid.*

Once your construction documents are ready and you've identified three contractors with whom you'd like to do business, it's time to invite them to a pre-bid meeting This gives them a chance to thoroughly review your construction documents and do a walk-through of your home for context. While some homeowners do this one-on-one, I prefer having one meeting attended by all of the contractors, along with their subcontractors. This way they'll see and hear the same presentation. Otherwise, the inevitable story variations that occur one-on-one could skew the bids. As a courtesy, let them know that this is a competitive bid and that other contractors will be in attendance.

MONICA'S TIP:

Having the contractors see who else is bidding on your project will undoubtedly get them to realize they're in competition for your business. The caveat is that because of this, a contractor may low-ball your project because they think you perceive remodeling contractors as a commodity. They'll over-promise on price or schedule with the intention of under-delivering on quality. If this happens, you'll need to review their bid with a fine-tooth comb to uncover why a bid is so low.

PRE-BID MEETING
AGENDA

INTRODUCTION & PROJECT OVERVIEW
Welcome the contractors and thank them for their interest in your project.

PLANS & SPECIFICATIONS
Distribute plans and specifications. These items should be printed, collated, and bundled as a bid package.

Bid Package: (List what is included in the package.)

—   —
—   —
—   —

SCOPE OF WORK
Present overall scope of work. The designer on your team lays it all out for the contractors.

■ **DISCUSS INDIVIDUAL PLAN SHEETS.** (The design pro presents these in detail.)

Plan Sheet 1:   Plan Sheet 4:

Plan Sheet 2:   Plan Sheet 5:

Plan Sheet 3:   Plan Sheet 6:

■ **DISCUSS GENERAL NOTES & INSTRUCTION SHEETS.** Again, leave this discussion to the design pro.

PROJECT WALK-THROUGH
Lead the contractors through your home. Since most of the questions will be related to the construction documents or the actual construction, let your design pro take the lead here.

BID DEADLINE
Give the contractors a specific date that is **two to three weeks out**. Let them know how long it will take for you to review the bids and make a hiring decision.

Deadline Date:

Hiring Decision Date:

Adjourn Meeting:
Thank the contractors for coming and reiterate that they are instructed to bid solely on the written documents and not on any verbal information, and that you look forward to receiving their bids.

ADJOURN MEETING

MONICA'S TIP:

Require that all subcontractors attend your pre-bid meeting. This will ensure bid accuracy and avoid surprise cost overruns. Stick with the agenda and allow others to share their suggestions, taking note. If you intend to incorporate any contractor feedback that will affect the documents, hold off making those changes until you've received and reviewed all the bids. After that, any changes should be documented and the contractors invited to adjust their bids accordingly.

Preparing a bid takes time because the contractor and subcontractors need to determine whether the construction documents contain anything unusual that requires special pricing. These factors could include difficulty accessing your property, limited space to store materials or move vehicles on the site, and the limited availability and/or high cost of materials.

The contractor will factor in the requisite labor and material costs for each phase of the job, adding a percentage for overhead and profit. This information is carefully prepared and presented in bid form. The bid should reflect everything on your construction documents and also outline change order pricing.

Your bid will be derived from a combination of subcontractor quotes, quantity takeoffs, and what we call "construction means and methods." Let's examine each of these pricing components in detail.

Subcontractor quotations are the subcontractor estimates for work that will not be performed by your general contractor. This can include electrical, plumbing, HVAC, lighting, cabinets, countertops, roofing, windows, and other specialties. Subcontractors should visit your home to compare construction documents to the actual site. It's important to note that general contractors do not control the subcontractors' pricing. However, the general contractor can use a competitive bidding approach in which several qualified subcontractors submit fixed-priced bids, with the general contractor awarding the contract to the lowest bidder. Alternatively, the general contractor can negotiate with one preferred subcontractor or even do a combination of the two—a competitive bid for run-of-the-mill tasks such as demolition, and a negotiated bid for specialized tasks such as custom cabinetry and crown molding).

Quantity takeoffs refer to a more detailed estimation process that accounts for all materials and labor required for your project. For example, the amount of wiring needed for electrical work, the number of door handles, plus the amount of the associated labor involved. The general contractor (or a member of his or her team) will create a forecast of what your project will really cost based on your construction documents.

Construction means and methods is a phrase that gets to the nitty-gritty of putting your plan into action. It's the series of activities that take place to complete your project. When the professionals involved take a closer look at what needs to be done, what additional tasks will have to occur to complete the primary tasks, and what costs are added because of them. For example, will scaffolding be required? That adds time, costs, and perhaps extra personnel. Will minor adjustments need to be made in the placement of conduit, doors, or windows because of obstacles that were not immediately apparent? If so, those work-arounds may come with added cost. They are likely unavoidable and were not apparent until all the other pieces were known. But they are still real, and they'll affect your budget.

THREE BID TYPES: *Which is Right for You?*

There are three common bid types: cost-plus (a.k.a. time and materials), guaranteed maximum price, and fixed price (a.k.a. lump sum). There are pros and cons to each, and job types to which each approach is best suited.

1 *Cost-plus*

In a cost-plus bid, the contractor submits what he expects to be the wholesale cost of the job, and then adds a percentage for profit. While this fee is pre-determined, the costs are variable.

Making a cost-plus contract work will require a lot of oversight and will also require you to pre-define allowable expenses in order to ride herd on costs. Giving anyone a blank check encourages inefficiency, which in turn increases the cost of your project. So, you'll have to create strict parameters up front to constrain costs. You must also define reimbursable expenses in excruciating detail and decide in advance whether things like payroll taxes, equipment rentals, construction wages, and the upkeep costs of equipment the contractor owns and uses on the job are legitimately reimbursable. Contractors should not be able to shift normal overhead costs into your cost-plus arrangement, nor should they be able to leave you holding the bag if their negligence or the poor management of their subcontractors incur additional expenses or interest fees.

Pros: The final cost may be less than that of a fixed price contract because the contractor does not have to inflate the price to cover his risk. This may result in better quality because the contractor does not have to skimp on materials and labor to meet the pre-arranged price.

Cons: One word—uncertainty. This type of contract is usually a poor choice if you have a strict budget. There is less incentive to control costs, since you've agreed to pay whatever the expenses turn out to be. Also, this type of bid shifts risk onto the buyer because of the opportunities for a sloppy or unscrupulous contractor to dump in additional expenses. Contractors' estimates will likely be less than accurate since expenses are passed through to you. Lastly, some states deem time and materials contracts illegal for home improvement work because the contracts are required to list start and stop dates and a structured payment schedule based on work completed.

② *Guaranteed Maximum Price*

Under a guaranteed maximum price bid, the contractor agrees beforehand that the cost of the work will not exceed a specified figure, known as the guaranteed maximum price (GMP). The GMP is based on competitive bids from each subcontractor, and on the contractor charging an additional fee for taking on the risk of the guarantee. The contractor is also allocated some contingency to pay for construction changes that are within the design intent of your project. Changes beyond the design intent require prior approval from you, and the contractor collects an additional fee under a change order. A "savings clause" provides a bonus percentage to be paid to the contractor if the job comes in below the maximum allowable cost.

Pros: With the maximum price set, you are assured that enough money is on-hand to pay for the project. And if the job comes in below budget, you receive the difference (minus the bonus percentage paid to reward the contractor for good management).

Cons: If the job comes in well under the maximum price, the contractor may stand to make more profit by charging you the maximum, which may be more than the bonus percentage. This may create a disincentive to provide an honest maximum bid.

 Fixed-Price

Fixed-price bids can be set up to pay out when the project is finished, or on a negotiated payment schedule. I prefer this arrangement since paying as work is completed brings peace of mind to both you—because you want to see work getting done—and the contractor, because she wants to keep the cash flowing.

Pros: A fixed-price bid is the most common type of bid for residential remodeling. This type of contract shifts the risk back to the contractor, who must determine if she can complete the work for the agreed-upon amount. Using a savings clause encourages efficiency, rewarding contractors that save you money. A fixed-price bid gives you peace of mind because you've locked in your price, even, say, if the price of lumber rises.

Cons: If the price of materials drops suddenly, you end up paying the higher price. Also, if an unusual market situation results in a dramatic increase in the cost of materials or labor, the contractor may need to renegotiate the fixed-price or be unable to complete the project for the original bid, which may put her in the awkward position of walking off the job.

During the screening process, you should have attracted the highest-quality contractors. You goal now is to get them to offer you their best price. Don't base your decision wholly on the lowest bid. Quality, experience, and reputation may well legitimize a higher price and provide fewer job headaches and greater long-term satisfaction. Be especially wary if one bid is far higher or lower than the competition. Remember, your contractor is your partner in creating your dream remodeling project. This is the wrong time to be penny-wise and pound-foolish.

11. REVIEW THE BIDS

*Discover how to compare bids and
make an informed decision.*

Do you remember what it was like to buy your home or car? Part of your due diligence included visiting homes or test-driving cars to compare features. You'll want to perform this due diligence again when comparing construction bids. I'll show you how to assess the contractors' bids on an apples-to-apples basis that not only compares the total price, but also what's included in that price.

The best-case-scenario is that your plans and specifications are so accurate to your desired outcome that later change orders and revisions are minimal. That means all the estimates, prices, and computations must be verified so you don't run into problems (or cost over-runs) later. By now you've also asked the candidates to revise their bids based on any suggestions you are planning to incorporate from the pre-bid walk-through.

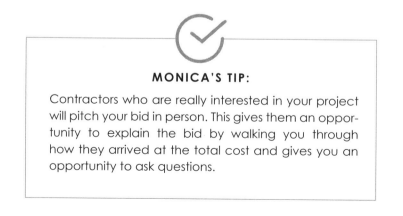

MONICA'S TIP:

Contractors who are really interested in your project will pitch your bid in person. This gives them an opportunity to explain the bid by walking you through how they arrived at the total cost and gives you an opportunity to ask questions.

Good communication with your contractor (and between your contractor and his/her subcontractors) is the most important way to reduce the need for changes later on. Your team needs to feel comfortable speaking up if they catch an error so changes can be made early in the process. Don't shoot the messenger! Be grateful if something is caught before it's too late. It's essential that the contractors and subcontractors be an active part of the pre-construction review so potential problems (or overlooked opportunities for savings) can be identified early.

One advantage of obtaining bids from more than one contractor is that you get a sense of their thought process, as well as how they view your project and whether they're able to add value with constructive feedback.

Once you've reviewed the bids' bottom line costs, take a second pass and pay attention to how specific and detailed they are. While remodeling is not rocket science, it does require the contractor to have the ability to see the big picture (i.e. an understanding of your vision to achieve the final outcome), as well as the ability to focus on how it will get done and which building materials to use. Whoever is estimating your project needs more than math skills. He or she also needs excellent listening, reading comprehension, strategy, visualization, and oral communication skills.

Quality bids identify the appropriate quantities of materials and labor required. They also reflect current material pricing since this can fluctuate depending on supply and demand or compressed delivery time. For example, if it normally takes ten to twelve weeks for a special order item and the vendor can produce it in six to eight weeks, there will likely be a premium charge for this. Find out up front what that charge would be and decide now whether you're comfortable paying that upcharge.

Avoid hiring a contractor that estimates too low. Some will intentionally low ball a job with an unrealistic but attractive price, counting on being able to later justify tacking on additional expenses. They know it's difficult and expensive to switch contractors in the middle of the job, and they're counting on the fact that you won't want the hassle. You also want to avoid contractors who are expert craftsmen but lack the business skills to estimate your project properly, including their profit. This type of contractor is prone to walk out on a project because they simply can't afford to stay on it.

One way to identify the low bidders is by checking references, reading Better Business Bureau ratings, and paying attention to comments by prior customers on sites like Yelp or Angie's List. Here's one place to trust your gut. If a price is significantly lower than competing bids without a clear reason, you're probably going to end up paying more in the long run.

That's why I always caution clients to be wary of any bid that's 10%–15% (or more) lower than their conceptual budget. When assessing bids, don't just focus on the total cost. Pay attention to the cost of each line item, the number and qualifications of subcontractors, and the quality of materials specified. You certainly don't want to splurge on your dream spa bathroom only to discover months later that you've had a leak caused by poor-quality materials or installation. It's only when you're paying for an expensive repair that you realize you've ended up paying more than having the work done right the first time.

When reviewing bids, keep your attention on the facts documented in black and white. If you find that contractors are requesting a lot of clarification or spotting perceived errors, take note and let the pro that prepared your construction documents know so the documents can be revised accordingly. Also note feedback or suggestions contractors may have. For example, a contractor may suggest reducing the number of ceiling lights in your lighting plan because he/she thinks the lighting plan is overdone. Ask why, but also consider that people are more prone to complain about not having enough light than having too much of it. Factor in the professional's suggestion, then make an informed decision that best suits your needs.

Earlier, I mentioned the importance of having the contractor actively engage subcontractors in bid preparation. This is because the subcontractors are more intimately involved with the work they'll be doing than the general contractor. So, for example, when reviewing the portion of the bid related to electrical work, pay attention to any differences in the bids and obtain clarification from the contractors about these differences.

I suggest framing the conversation along the lines of "Thank you for taking the time to prepare your bid for my project. When I was reviewing your bid, a few questions came to mind regarding the electrical aspect. I'm hoping you can help clarify a few things for me." Approaching the contractor from this perspective lets the contractor know you're receptive to listening.

When you review the bids, ask the contractor how the schedule was developed. More specifically, you'll want to know that she or he applied the critical path method (CPM) to figure out the amount of time needed to finish your project. CPM requires a contractor to list all the activities involved with your project along with the amount of time necessary to finish them, and then identify which activities are dependent on completion of the others before your project can move forward. Activities that must be done on time for your project to be completed on time are considered "critical." The CPM impacts how subcontractors are sequenced and makes it clear which delays will throw off the projected completion date.

Discuss the CPM with the contractor and find out where the vulnerable points might be. Are there rumored to be shortages or price hikes coming for key materials? Are there any big new local construction projects that might make it difficult to get specialized subcontractors at critical points? Ask the contractor how he/she intends to work around those issues if they occur.

We often say that you get what you pay for, but that's not always true with a high bid. If a contractor's bid is significantly higher than your conceptual budget, don't assume that a higher price means you'll get better quality or faster service. If you liked the high-bidding contractor, ask for a detailed explanation of why the bid was such an outlier and see if the answer justifies the price.

Evaluating bids is no easy feat since contractors may not present the same information in the same way. Commercial construction projects commonly utilize a specifications-writing standard established by the Construction Specifications Institute that classifies work by results, so the detailed information in your bid is organized. I suggest organizing the remodeling bids you receive in a similar format, starting with General Requirements.

MONICA'S TIP:

You know what they say…"If it's too good to be true…it probably is." So, before you decide to take the lowest bid, particularly if it is substantially lower than your conceptual budget/estimate or the other bids, be absolutely sure the contractor can deliver. How? By checking several references and asking whether their projects were completed within the amount specified within the original budget and time frame. If not, ask why their projects did not adhere to the contractors' originally submitted time, bid price, and schedule. If the reason the project was not completed on time or on budget had to do with the contractor's incompetence versus something that was out of the contractor's control, move on.

PARTS OF A REMODELING BID

The **General Requirements** section of a bid is typically reserved for non-direct costs such as architectural/engineering services, plan check, permits, trash removal, rental equipment, job toilet, miscellaneous, dust protection, finishes protection, supervision, general labor, liability insurance, and final clean-up. The next category of information to compare is **Site Construction**, a.k.a. site work. If there is any site work or landscaping to be done, it would appear here. **Concrete** is comprised of pre-cast concrete products delivered ready for installation as well as site-cast products. **Masonry** deals with materials such as brick, concrete block, glass block, stone, and tile. The **Metals** category ranges from structural steel framing to any other metal such as the metal roofing or stairs. **Woods and Plastics** also includes composites that are made up of more than one material to create better durability, such as composite decking.

Thermal and Moisture Protection covers any materials used to damp-proof or waterproof a home, with the exception of doors and windows. The **Doors and Windows** category covers any opening that provides access to your home or allows for air flow or light transmission. **Finishes** is the most exciting category because it deals with the visible items such as wood flooring or tile.

Specialties is a catch-all for items, such as fireplaces, that don't fit into other categories. The **Equipment** category refers to any equipment required for your project. In addition to furniture, **Furnishings** includes window treatments and casework such as cabinets. Items in the **Special Construction** category are typically those purchased pre-assembled in whole (rather than installed in parts) and installed by the contractor. A perfect example of this are She Sheds, a.k.a. woman caves.

Conveying Equipment covers any transportation within your project, such as dumbwaiters, elevators, and handicap lifts. The **Fire Suppression** category consists of items that prevent or extinguish fires. **Plumbing** includes any item that is part of the system that conveys fluid ranging from pipes and valves to the plumbing fixtures themselves. **Heating, Ventilation, and Air Conditioning (HVAC)** includes items related to the quality of indoor air as well as thermal comfort. The **Electrical** category is all about power and lighting.

Data, voice, and audio-video communications are included in the **Communications** category. The **Electronic Safety and Security** category includes electronic burglar alarms and the accompanying hardware, software, and equipment; controls for electronic access such as security gates; closed circuit video surveillance equipment; and alarms that notify you of the presence of fire, smoke, radon gas, and carbon monoxide.

Earthwork involves backfilling, compacting, excavation, grading/leveling, and/or demolition, as well as disposal of excess earth and debris. The Utilities category refers to electricity, water, gas, sewer, etc. Lastly, Overhead and Profit is focused on the contractor's operating expenses for equipment and facilities (overhead) and the difference between the cost of goods and the price for which they are sold (profit). Overhead includes expenses such as office rent and utilities, licensing, salaries and benefits for employees, and advertising.

Once you've reviewed the applicable categories for each bid, it's time to conduct an apples-to-apples comparison. Use a spreadsheet comprised of three columns (one for each contractor) and enter each line item under the appropriate category.

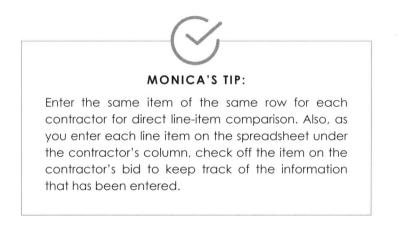

MONICA'S TIP:

Enter the same item of the same row for each contractor for direct line-item comparison. Also, as you enter each line item on the spreadsheet under the contractor's column, check off the item on the contractor's bid to keep track of the information that has been entered.

Once you've completed the data entry, ask these initial questions about how the estimates compare:

- Are they close? If not, how do they vary?
- What's missing in the lower bids?
- What's included in the higher bids?

Some differences may be the result of contractors not capturing the full scope of work, allocations for finish selections that are unrealistically low, difference in contractors' quality expectations, different specifications, and/or lack of plan detail.

MONICA'S TIP:

Don't forget about the schedule's impact on the bid. For example, a bid with a shorter completion time may be priced higher.

There are no shortcuts to reviewing and assessing bids. Take the time to really understand the similarities and differences between them. Only then will you be able to identify the bid(s) that most accurately reflect the scope of work presented during the pre-bid meeting.

12. GET THE MONEY

*Learn the best way to get
all the money you need.*

You should be proud of the work you've done thus far! You've taken a hard look at all of the costs involved with your project, including a contingency, and have a good handle on how much money you need. Sanity-checking your finalized remodeling budget against the solid bid in hand is the critical indicator that you're ready to apply for financing or set aside the funds from your home equity line of credit, or savings. In addition to the contingency figure in your construction budget, it's a good idea to apply for 10%–15% more than you think you'll need. Consider it an extra contingency cushion. If you don't need it, you can use it to pay back your loan faster.

Applying for a loan is so much more than filling out an application. It's telling a story about your credit history, income, and the cash flow necessary to cover loan payments. Realize that you and the financial organization have different objectives. You are focused on getting the money to achieve your dream. The lender wants to make money in interest, while minimizing the risk of not getting repaid or making less on your loan than on someone else's. Be prepared to provide lots of documentation about your project and your solvency. You'll be glad you have detailed, solid documents to back up your request for cash.

MONICA'S TIP:

When it comes to condos, lenders want to see that the complex is occupied by a high percentage of the owners of the units and not renters. They don't want to lend on one of only a handful of units that are not rentals because a condo complex with a high percentage of rentals is viewed more like a hotel or timeshare and thus considered more risky. You'll also want to show the lender that your complex is well managed, with the homeowner association's reserves meeting the minimum required by your lender so cash is on hand for major repairs and expenses.

CHOOSING A LOAN ADVISOR

There are a lot of mortgage loan products out there, and it makes sense to have someone on your side who can sift through these loans based on your credit score and other factors to match you with the best loan (rate, term, and costs) for your needs. That someone will be a loan advisor (a.k.a. loan officer or a mortgage broker). While their ultimate goal is to get you financed, not all of them work the same way.

Loan officers work for a specific lender, have access only to loan products offered by these institutions, and are paid by them. Mortgage brokers, on the other hand, develop relationships with more than one lender. As a result, they will not have access to every loan product that may be a good fit for you. Also keep in mind that lenders pay mortgage brokers a commission based on a percentage of the loan. So, you'll want to work with an ethical one who has your best interest at heart and conducts business with transparency, disclosing their fees prior to submitting a loan application. That said, working with a mortgage broker is typically more advantageous than working with a loan officer who is only able to present you with loan products offered by their employer.

While they are not mortgage brokers per se, companies like Bankrate.com and NerdWallet are also great for researching mortgage rates from multiple lenders. By entering information anonymously in the loan comparison tools on their sites, you're able to check out several lenders and compare what they have to offer. Another option is to connect with an online lender like SoFi that looks beyond a three-digit credit score to holistically evaluate your application, because you are more than your credit score.

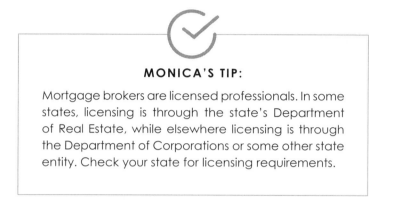

MONICA'S TIP:

Mortgage brokers are licensed professionals. In some states, licensing is through the state's Department of Real Estate, while elsewhere licensing is through the Department of Corporations or some other state entity. Check your state for licensing requirements.

APPLYING FOR THE LOAN

At this point you've done three things to ensure that your money doesn't run out during construction:

1. You have a well-thought-out remodeling plan that accurately reflects your wishes.

2. You've developed a realistic budget based on a carefully prepared bid.

3. You've asked for 10% to 15% more money than your bid number and contingency.

If you haven't already locked in your loan, make sure the loan types and terms haven't changed since you did your research. Be open to new options that might have been added in the meantime—there may be a better deal available. Thoroughly discuss the pros and cons of any new type of loan with your loan advisor to make sure it's as good as the original plan. Be wary of loans that offer big savings now and balloon payments later, since your income may not rise as quickly as you hope, or unexpected expenses down the road may make those larger payments burdensome.

It's worth checking to see if there are any new incentives available to sweeten the deal. These might be from your lender or from the government. Look for new programs that encourage energy efficiency, historic preservation, reduced environmental impact, etc., as they apply to your project.

Ask your loan advisor and your general contractor if they've heard of any new incentives and do a little research online, to be on the safe side. Even if the incentive saves you only a few hundred dollars, it's money in your pocket for free!

If you identify a new loan option or a new incentive, make a thorough comparison to ensure that you're not trading off some benefits for others or incurring a hidden cost. Check the interest rates, payment plan, and terms/conditions for any hidden surprises. Watch out for time limitations, additional documentation, or new restrictions. If necessary, have a lawyer and/or your accountant consider tax or income ramifications.

Organizations offering incentives always realize some benefit to themselves, so make sure you know what's in it for them before you jump on a fantastic deal. Don't be afraid to sweat the details and have your loan advisor explain the nitty-gritty. If an incentive or loan product is new, do some research to see what other experts and users are saying. It's usually best to avoid overly complicated financial maneuvering, and to run away from something that seems too good to be true.

Be prepared to submit a Letter of Explanation with construction documentation, if asked. Lenders want to know that you're well-prepared to embark on a remodeling project. They also want assurance that your project is going to add value to your home. The ability to provide this information in a timely manner will signal that you've done your homework.

Finally, keep your contractor informed of the finance timing so he or she will be prepared to start as soon as you give the go-ahead. However, remember that money talks. Your contractor will likely give priority to the homeowner with financing in place.

CLOSING THE DEAL

Once you fill out a loan application in person or online, the lender will assign a loan consultant to your case. He or she will guide you through the process and let you know exactly what documentation you need to provide. Resign yourself to reading a lot of documents that will require your signature.

The law requires lenders to provide a good faith estimate on the loan amount and terms no later than three days after the application is complete. This document will include closing costs and other fees, and a truth-in-lending disclosure that spells out the terms, finance charges, annual percentage rate (APR) of interest, and monthly payment schedule. Since interest rates change every day, the document will also tell you how long you're guaranteed the stated rate. If there's a balloon payment or pre-payment penalty, it should also be in the good faith estimate.

Once your documents are signed, they'll be turned in for underwriting review. The lender will verify your income, employment, and financial profile. A professional appraiser will be retained to appraise the property. The appraised value will need to be as much or more than the purchase price if you are buying the home. If you are remodeling a home you already own, the appraisal will help the lender determine how much money they are willing to loan.

Stage your home to maximize the lender's appraisal. Consider quick, easy, and inexpensive ways to spruce up curb appeal and stage your interior. Walk through your home with the appraiser. Find out how long she has been in business and which neighborhoods she is intimately familiar with. Show the appraiser information about comparable homes that have sold in your neighborhood. If you've visited any of these homes, share any differences and/or similarities that will help your home appraise for top dollar.

The lender will guide you through other necessary steps, like obtaining flood certification and title insurance. The costs for these items should be detailed in your closing costs. The institution will ensure that you have homeowner's insurance as a condition of issuing the loan.

You're not done yet! All of that data goes to the back office for further review. If there are problems or missing items, the loan will be flagged and your consultant will let you know what's needed. Once the loan is approved, you'll set up a closing date to sign the loan documents. In the meantime, the lender will make sure your title to the property is clear, meaning your ownership is unquestioned and the property is unencumbered by liens. The institution will also look over your homeowner's policy to make sure it is sufficient to protect the property.

Before your appointment, your consultant will give you the final closing costs and let you know what to bring to closing. At a minimum, you'll need proof of your identity plus a cashier's check for the closing costs. If you're buying a home, you'll need another cashier's check for the down payment. If you're getting a loan through an online lender, they'll have you upload the requisite documents and transfer closing costs.

Allow time at closing to read over the documents. The stack may be close to an inch thick! You have a right to understand everything. If something seems off compared to prior discussions, now is the time to ask questions. If you find an error, say something. It's much better to have a small delay while paperwork is corrected than to try to deal with an error once everything is finalized. You should leave with a copy of the documents for your files.

If you are purchasing a home in this transaction, you'll receive the keys at closing. If you are refinancing, you leave with documents and a loan. Depending on the state, there may be a several-day right-of-rescission period during which you can cancel the transaction. If you've carefully done your homework, worked out your finances, and projected your costs, nothing about the closing should come as a surprise, so there should be no need to exercise that option.

You and the loan specialist will discuss payment milestones based on completion of building phases as well as the request for the initial disbursement of funds. It's common practice for the bank to convey the money in the form of a two-party check that must be endorsed by both you and the contractor. Generally, you'll receive the initial check after all of your permits are in place. Even though you select the contractor to do the work, the lender will want to confirm that the contractor is qualified for your type of project.

Expect that either the permitting process and/or the lender will place a time limit by which the renovations must be completed. So, keep on top of the renovation schedule to ensure your project is not delayed. Your contact person at the bank will call or email you periodically to get a progress report. This person may even visit the job site to see first-hand evidence that work is progressing according to the construction schedule, and will expect an explanation if the project is not moving forward as planned. You and your contractor should be mindful that draws will be disbursed according to the lender's progress payment procedures.

Once the remodel is completed, your contact person at the bank will work with you to request final disbursement. Banks and states may vary in their requirements. Your contractors will need to sign a lien waiver that assures the lender there are no potential mechanic's liens (a legal claim on your property for non-payment of subcontractors and materials suppliers) pending or already filed against the property. A final inspection will be scheduled, and the results will need to meet the bank's approval. Only then will the final check be cut, and once again, it will be made out to you and the contractor, requiring both signatures.

13. REVIEW THE CONTRACT

*Learn the elements of
a solid contract.*

A contract is a written agreement between you and the contractor that thoroughly describes the services to be performed, the labor and materials to be furnished, and how much it will all cost. A contract also describes when and where the work will be performed. Your contract should include every single thing you've said you wanted. If you're expecting a certain style, make, model, color, size, quality, etc., make sure it is explicitly referenced in the contract. This way there will be no room for misinterpretation. Operate from the standpoint that if anything you discussed is not written, it's not real, and therefore the contractor is not responsible for doing or providing it. Did you want to donate your old cabinets to charity? Make sure it's in writing. Does your homeowners' association have requirements about the placement of dumpsters or where workers can park? It should be noted in the contract. Be sure you cover details like the clean-up inside and outside, who arranges for the waste removal, any special accommodations for pets or children, and other details.

MONICA'S TIP:

Ask yourself—if I gave this paper to a stranger, would he or she get everything exactly right? If not, go back and add more detail.

Check with your state about the information required to be included in your contract, since the requirements vary from state to state. At the very least, make certain your contract includes these key pieces of information:

1. **Name and address of contractor:**
    Where you would send any notices in writing to the contractor.

2. **Name and address of homeowner:**
    Your mailing address, if different than the project address.

3. **Name and address of the project:**
    Where the work will take place.

4. **Name and address of the construction fund holder:**
    The entity holding the funds if they were not disbursed to you directly.

5. **The price of the work to be performed by contractor:**
    The grand total.

6. **The amount of the finance charge (if any):**
    If your contractor offered financing for a portion of your project, the finance charges and how they are calculated should be detailed here.

7. **Project description and detail on the equipment and key materials involved:**
    This is the narrative that helps set expectations.

8. **Property lines:**
    It's a good idea to give the contractor a map of your property, confirmed by a licensed land surveyor.

9. **Down payment:**
    Should reflect an amount not exceeding the maximum allowed by your state.

10. **Contractor's license number:**
    Be sure you've verified that it's active.

11. Schedule of progress payments:
How much and when to pay the contractor, including a retention (more on this later).

12. Approximate start date:
When substantial work is expected to start and what constitutes substantial commencement.

13. Approximate end date:
When to expect substantial completion.

That's just the beginning. Now for the various clauses:

1. Delay:
Defines criteria that constitutes a delay.

2. Mechanic's lien release:
Defines the type of mechanic's liens and when they are due.

3. Plans, specifications, and permits:
Specifies which plans and specifications will be used for your project and that the contractor is responsible for obtaining and paying for permits, while you are responsible for paying for public agency or utility assessments and charges.

4. Extra work and change order:
Defines what constitutes extra work and how changes should be documented.

5. List of documents incorporated into contract:
Calls out any documents not already alluded to in the contract.

6. Labor and materials:
Specifies that the contractor is responsible for paying for these items unless your payments to the contractor are in arrears.

7. **Contract, plans, and specification:**
   Provides guidance on what to do if these items that are intended to supplement one another are the source of conflict, with the plans having priority over the specifications and the contract provisions controlling both the plans and the specifications.

8. **Allowances:**
   A placeholder for the estimated cost of finish materials, such as tile, that have not been identified (only if absolutely necessary).

9. **Completion and occupancy:**
   Once your project passes final inspection, the contractor should have received all payments, excluding retention, and will record a Notice of Completion within the necessary time frame.

10. **Destruction or damage insurance:**
    Certifies the type of policy the owner has in place and excuses the contractor from performing further work on your property as a result of events outside human control, such as fire, flood, and earthquake.

11. **Right to stop work:**
    Gives the contractor the right to stop work if not paid.

12. **Limitations:**
    Provides a deadline for filing a lawsuit after completion or stalling of work.

13. **Attorney's fees:**
    States what types of costs and expenses are awarded to the prevailing party.

14. **Clean-up:**
    Identifies who is responsible for cleaning up once work has been completed and defines what constitutes clean-up.

15. **Notice:**
    Provides instruction on how to give notice.

16. Prohibition of assignment:
Requires written consent from the owner for the contractor to assign the contract or payments to another party.

17. Bankruptcy:
Gives grounds for either party to cancel contract if either party becomes bankrupt.

18. Commercial general liability:
Specifies the company that wrote the contractor's commercial liability policy and also provides contact information for the insurance.

19. Workers' compensation insurance:
Certifies that the contractor carries this type of insurance for employees.

20. Arbitration:
Requires you and the contractor to resolve disputes through arbitration.

21. Performance and payment bond:
Should you require them, certifies that the contractor will perform per the terms and conditions of the contract, and also certifies the laborers, material suppliers, and subcontractors against non-payment.

22. Mechanic's lien warning:
Describes how to prevent mechanic's liens.

23. Three-day right to cancel:
Protects you from the pressure of in-home negotiations. If the contract is negotiated at the contractor's office, this clause does not apply.

24. Warranty:
An assurance that the contractor will stand by his or her work.

MONICA'S TIP:

A written contract (preferably in plain, easily understood English) is required, no exceptions. Do not rely on a contractor's reputation and a handshake. Make sure that all sections of the contract are filled out. If a section does not apply, it should be lined through and initialed or clearly marked "not applicable." Be certain that you understand every detail and clause in the contract before you sign. Have a lawyer look it over if you are unfamiliar with contracts and legal language. Once you sign, it's difficult to make changes, and you're dependent on the good will of the other party to agree to modifications. You and the contractor should both keep paper copies of the signed contract in case questions arise later.

WARRANTIES

The terms of a warranty spell out when it starts and when it ends, what's covered and what's not covered, any conditions that may void coverage, who to contact for service under the warranty, and whether the contractor will repair or replace a failing material or whether he/she will refund your money. You'll also want to obtain the written manufacturer warranties for appliances or other materials installed by the contractor (even for items you can't see, such as insulation).

Even the best building and finish materials can fail when installed correctly. Certain factors—what they are made of, how you use them, and under what conditions—have a dramatic impact on their durability. While you are unable to predict whether you're going to have problems, be aware of the reliability norm for the building and finish materials. Take the time to research products and materials that undergo extensive unbiased physical testing and are rated by independent organizations such as Consumer Reports.

Here's the scoop on two types of warranties that should be included in your contract:

Implied Warranties

State laws govern implied warranties, those that exist even though they are not expressly written into your remodeling contract. For example, it is implied that the contractor is going to comply with building codes and perform your remodel in a workmanlike manner that befits the minimum standard of quality. In other words, your home should be habitable once the work is completed. Believe it or not, you actually imply a warranty of constructability (also known as fitness of plans and specifications) in which the plans you've provided to the contractor are complete and fit to build your intended project. Check with your state to verify what other implied warranties there may be, if any, for your remodel.

Expressed Warranties

An expressed warranty is written out. It can be stated verbally, though it's hard to prove when contested, so always get it in writing. The three types of express warranties you should include in your remodeling contract are a Fit and Finish warranty, a Quality Work warranty, and a Defect-Free warranty.

The fit and finish warranty is a limited warranty that covers the fit and finish for cabinets, mirrors, flooring, interior and exterior walls, countertops, paint finishes, and trim.

It does not cover ordinary wear-and-tear or the normal effects of settlement, expansion, and shrinkage. For example, it wouldn't cover hairline cracks in concrete since this is not unusual. On the other hand, if you've found a severe crack you can put a penny through, bring it to the contractor's attention. Another example of something that would not be covered is a flaw caused or made worse by you or your guests.

MONICA'S TIP:

Always follow the maintenance terms of an express warranty. Failure to do so will void the warranty.

What would be considered a fit issue versus a finish issue? Defects in materials or workmanship that are not part of generally accepted installation and quality standards for your state are a fit issue. The installation and quality standards need to be viewed locally, since what's generally accepted for mild- or low-humidity climates may not be the same for extreme or high-humidity climates.

Finish-related issues are often visible smudges, scuffs, scratches, chips, and stains not attributed to minor imperfections. It's like the disclaimer you see on items made from natural materials that lets you know the imperfections are a natural part of the product. Examples of finish-related issues range from a chip in the tile to missing or torn window screens, broken or cracked glass, scratched tubs, and missing items. You'll know it when you see it.

MONICA'S TIP:

Minor adjustments are inevitable. Unless it is an emergency, live in the house for a week or two and make note of any adjustments that may need to be made to doors, windows, cabinets, toilets, plumbing fixtures, etc. Collecting a list of adjustments will show that you value the contractor's time so he or she doesn't have to make several return trips. Tell the contractor about the problem in writing as soon as you identify it.

The Quality Work warranty covers damages linked to installation errors. The Defect-Free warranty relates to structural elements such as the roof, walls, and foundation. While the first two warranties are good for a year after the municipality issues a certificate of occupancy, the Defect-Free warranty extends for a longer period. This is comforting because structural issues resulting in a sagging roof or floor joists could take years to be discovered. Check with your state to confirm the length of the Defect-Free warranty where you live.

FILING A WARRANTY CLAIM

If you need to file a claim after your project is completed, notify your contractor promptly in writing. The contractor is then allowed to review your claim and offer to repair the faulty work or settle the claim. If the contractor does not resolve it within a reasonable time frame, you have the mediation, arbitration, small claims court, and hire-a-lawyer options presented in chapter eighteen. Another option is to contact the state agency that licenses and regulates contractors by the deadline. You may want to start with this option first so a consumer advocate can help you better understand your rights and offer guidance.

Everyone hopes for a problem-free remodeling project. In many if not most projects, problems that do occur are eventually resolved. Knowing your options, rights, and responsibilities empowers you to retain control of your project and your time.

14. REDUCE RISK

*Learn five strategies that
will save you loads of heartache
in the future.*

No remodeling project is risk free. Therefore the potential for risk should not be ignored. As the late British Prime Minister Benjamin Disraeli once said, "I am prepared for the worst, but hope for the best." That's a sound mindset when you're heading into a remodeling project. Learn the tricks of a master money minder to keep your budget from getting away from you, and discover ways to reduce your construction costs while also learning where flexibility matters.

THINK OF YOURSELF AS A CEO

A CEO is a leader responsible for shaping and communicating the company's vision. This, in turn, determines the strategic direction for the company, and every decision ties back to this vision to keep the company on course. As a homeowner, you are very much like a CEO. Take an active role by keeping your fingers on the pulse of your project to ensure it's shaping up as expected.

CREATING A CLEAR CHAIN OF COMMAND

Good communication is key to a successful remodel. Early on, you defined the role and scope of work for each member of your project team. Before signing a contract with a contractor, you'll want to have a clearly defined chain of command once construction is underway. This includes a clear communication chain for input and distribution of project information. To do this, establish lines of authority and decision-making so it's clear who is accountable for what. Here are a few questions that will help define your chain of command:

- Who is your key contact if questions arise, either your own questions or those of someone on the crew or team if the boss isn't around?

- In what format will questions and other information be addressed?

- What happens if something does not go according to plan?

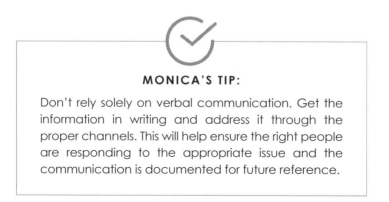

UNDERSTANDING INSURANCE AND BONDING

When homeowners think of risk, they often think of homeowners' liability insurance or property insurance. However, contractors are also required to have general liability insurance. This type of policy coverage pays in the event something becomes damaged, or worse, if someone becomes injured as a result of the contractor mismanaging a situation. It may also provide automobile liability coverage. Contractors should also have workers' compensation insurance that covers their employees' injuries on the job. It's important that you verify what the contractor insurance requirements are for your state and/or local municipality, and what's covered under the contractor's policy. Then, confirm whether the general contractor has actually read his/her policy and carefully examine your general contractor's certificate of liability insurance to confirm the necessary coverage is in place. If these safeguards are not in place, respectfully let your contractor know that you will be unable to take the risk of hiring him or her.

MONICA'S TIP:

Prior to commencing construction, have the contractor list you as a certificate holder on the certificate of liability. That way you will automatically be notified whenever the policy is updated, including if the policy lapses.

Don't assume your own homeowners' insurance is set up for the variables that can happen during major construction. Talk to your insurance agent about the work you're having done and ask whether you need to make any adjustments to your existing policy for the duration of construction. You may find that once the work is completed, you'll also need to adjust your coverage (since the property value will likely increase).

With a parade of people in and out of your home, it's unlikely you will be able to stake out the door and watch everyone come and go. They'll be parking in your driveway and on your street, walking on your sidewalk, using power tools in your garage. There may temporarily be holes in your roof or walls, or window openings covered by plastic sheeting or tarps. That's why it's important to discuss coverage pertaining to loss from fire, acts of God (such as flood, high winds, tremors/earthquakes), vandalism, removal of debris, and erection of temporary fencing/buildings, etc. Asking first and finding that your policy has you covered is a lot better than assuming and discovering too late that you're not covered.

Performance and payment bonds are also ways to mitigate risk. Performance bonds guarantee that a contractor will adhere to the terms and conditions of a contract. The hired contractor submits this bond. A payment bond guarantees that a contractor will pay suppliers, laborers, and subcontractors (subject to contract terms) for labor and materials. Generally, these bonds are issued together since they are closely related. However, if a contractor has had any bankruptcies, tax liens, or poor credit, he or she may be unable to obtain a bond. In many states contractors must show proof of bonding before a license can be issued or renewed. Contractors apply for these bonds, similar to how you would apply for homeowners' insurance. If your contractor is not bonded, that's a red flag and good reason to re-evaluate your choice of contractor.

MONICA'S TIP:

Be wary of a Blanket Performance and Payment Bond, one bond covering the value of *all* contracts a contractor may have open at any one time. These bonds are rare, but they do exist, and they tilt the contract very much in the favor of the contractor since the *entire* amount of the contract can be asked for up front and a mechanic's lien notice and payment schedule are not required. Most contractors are not eligible to post this type of bond since they don't meet the stringent requirements, particularly having a net worth ten times the total value of the bond. Because these types of bonds are so rare, you'll want to obtain written proof of its existence, effective dates, and claims history from the surety company. Contractors who qualify for blanket performance and payment bonds typically do a high volume of work. As a result, it's probably safe to assume they are in demand and do quality work. If you move forward, make certain that you fully understand the constraints and liabilities, and take a hard look at whether or not the situation presents you, the homeowner, with any benefits that outweigh the risks.

DO MORE THAN MITIGATE RISK . . . PREVENT IT

While insurance and bonds are ways to mitigate risk they don't prevent it. The most effective way to prevent risk is by ensuring that your general contractor and subcontractors conduct their business in a professional manner. Your team should follow accepted construction and business standards in the pre-construction phase, construction methodology, attention to safety precautions, and hiring team members with relevant industry experience.

BE THE ULTIMATE BEAN COUNTER

The term bean counter is synonymous with an accountant focused on controlling expenses and budgets. Let your bean counter alter ego track costs. You carefully crafted a budget as a financial road map; now keep an eye on the map as you move forward. Set milestones during construction to verify whether the costs match the amount of work that has been completed. Doing so allows you to catch discrepancies early so adjustments can be made, if necessary, long before completion. Driving is a good analogy: while a poor driver might only focus on getting from one intersection to the next, an excellent driver will constantly monitor the environment and anticipate the need to change lanes when new conditions first come into view.

MONICA'S TIP:

Being watchful gives contractors extra motivation to also keep a close eye on things.

If all is well, you gain peace of mind based on real-time information. When things don't go as planned (take this as a given), your budget will help you identify deviations from your financial plan. If caught early enough, it's often possible to make up slight budget or timing overages.

Likewise, if you are under budget in some areas, you can afford to splurge in other areas. For example, one of my clients was able to upgrade from French doors to a folding door system that made her project extra-special by creating a wonderful indoor-outdoor connection between the dining room and deck. Imagine having enough cost savings to splurge on that high-end appliance or other object of your desire. That's one of the potential benefits of keeping a close eye on your budget.

15. PAY SMART

*Learn how to avoid the
number one error people make
when paying for work.*

Believe it or not, remodeling homeowners were once accustomed to paying one-third down, another third when the job was one-third complete, and the final third upon completion. This did not serve them or the contractor well, because the homeowner was paying for work far ahead of completion and had no leverage if an issue arose, and the contractor was financing a considerable amount of the project. That's why it's important to set up a payment schedule in which you are paying for what's been done, while providing the contractor with a comfortable cash flow. The two of you will negotiate the payment structure based on milestones or dates, how much each payment should be, how many payments are due, and the completed steps between payments

I recommend basing your payment schedule on milestones you can see, such as the completion of site work, foundation, rough shell, and finish shell. I also recommend holding back a percentage of the payment until a final punch list walk-through has been completed. This is established industry practice. Reputable contractors should not take offense, and may even encourage the practice to ensure issues that come up are addressed right away.

Let's take a look at these suggested milestones, keeping in mind that your payment schedule may vary depending on the remodel's scope and local code requirements.

SITE WORK

Milestone **1**

Site work complete. Site work consists of non-structural tasks such as clearing the lot, digging the foundation, grading, installing temporary fencing, laying pathways for access, demolition, and installing temporary utilities.

FOUNDATION

Milestone **2**

Foundation complete. If you are adding square footage, you'll need a new foundation for the addition or to shore up the existing foundation to accommodate a second story. Pay for the foundation work after it passes inspection.

ROUGH SHELL

Think of the rough shell as anything that will be concealed in the walls, such as the framing and work done behind the walls—plumbing pipes, HVAC ductwork, and electrical wiring.

Milestone **3**

Framing 50% complete. Pay for framing, roofing, and framing materials.

Milestone **4**

Rough plumbing, electric, and HVAC 50% complete. Pay for plumbing, electrical, HVAC, and exterior siding. Pay balance due for framing based on passing inspection.

Milestone

Rough plumbing, electric, and HVAC 100% complete. Pay remaining balance for plumbing, electrical, and HVAC after they pass inspection.

FINISH SHELL

Once the activities associated with the rough shell have been completed, the walls, floors, and ceilings can be insulated, enclosed with drywall, and readied for finish materials.

Milestone

Drywall 50% Complete. Pay for insulation and drywall materials and labor based on progress.

Milestone

Drywall 100% Complete. Pay drywall balance after it passes inspection. Pay for interior carpentry and interior painting.

Milestone

Interior Carpentry Complete. Pay for interior carpentry, painting, and tile installation when complete.

Milestone

Cabinet and countertop installation, tile, grading, slab, asphalt, flatwork, and landscaping. Whether your cabinets are custom-built or prefabricated, installing them correctly is a big job. That's also true for countertops. Your cabinets will bear a lot of weight, and your countertops will see a lot of use, so make certain they're done well.

Cabinets should be the color and finish specified, and hung evenly at the proper height. Doors and drawers should be even and work smoothly. Shelving should either be installed or easy for you to customize with pins and pre-cut shelves. Trim, glass inserts, and door/drawer pulls should match your specifications. Cabinets should work with the specified under-cabinet lighting or appliances and not conflict with the placement of outlets and switches.

Countertops should be the color and composition promised, with the correct edge finish. They should be free of dings and defects and should sit level on the cabinets. Any sealant or protective coating should be applied and the finish should be satisfactory. The countertops should properly fit atop the cabinets and fit right around sinks, appliances and other built-ins.

Milestone 10

Finish plumbing, electrical, and HVAC based on passing inspection.

Milestone 11

Flooring materials and installation, and any other unpaid invoices.

Milestone 12

Pay the retention—percentage of contract value—when all the work has been completed under the terms of your contract.

MONICA'S TIP:

Throughout the job, trust that the general contractor you carefully selected will do right by you. However, never make the final payment before the final inspection has occurred, the inspector has signed off on the building permit, your punch list has been addressed, and you are satisfied with the work.

While you can quickly see when an activity such as demolition has been completed, it's not always easy to see 50% completion for framing or electrical work. How do you judge a job's progress? One way is to compare the amount of labor and materials applied to your project thus far to the amount required to complete it. If the labor and material costs are roughly equal to 50% of the completed value, you can presume 50% of the work has been completed. Luckily, most of the payment milestones are straightforward since they are based on work 100% completed and passing inspection.

Withholding payment is your biggest leverage to make certain you get the remodeling job done just the way you want it. Sure, you've got other options like liens and lawsuits, but those are messy, adversarial, and expensive. Keeping payments on schedule and linking them to work completed is the easiest and best way to maintain a good relationship with your contractor *and* get the outcome you want.

PAYING THE BILL

MONICA'S TIP:

Don't wait to unload your frustration with the contractor on payday. This will be viewed as a stall tactic.

Now that the time has come to pay, stick to the schedule. If change orders or unavoidable material or manpower delays have affected the pay structure, discuss the changes with your contractor. Consider breaking a large milestone payment into two or more smaller ones so you can pay for what has been done without getting ahead of the progress, thus preserving the interests of both you and your contractor.

Before making a payment, you should receive an invoice from your contractor. Make sure you keep a copy of the invoice and store it with your contract. The invoice must be in writing, and must provide details about:

- The specific work it covers
- The period the work was performed
- How much money is duev
- When payment is due

- How the contractor calculated the amount of the progress payment
- How to pay
- What will happen if you don't pay the amount invoiced

MONICA'S TIP:

Pay the same designated person each time and don't make the last big payment until after the local municipality has issued a Certificate of Occupancy. Without it, you're unable to use your home as planned.

If you disagree with the amount invoiced, notify the contractor within the time frame proposed in your contract and show your method of calculation for the alternative amount. Don't immediately assume fault or ill-intent. It may be that the invoice is actually correct and that you and your contractor are not communicating effectively. Or it might be that the invoice is wrong, but contains an error the contractor is willing to fix. Be firm and direct, but begin your inquiry giving everyone the benefit of the doubt and hone in on the details.

MONICA'S TIP:

Watch out for red flags such as a subcontractor or vendor asking for payment directly from you because the contractor's payment to them is long overdue. If this scenario should arise, bring it to the contractor's attention immediately so it can be taken care of. Another red flag is the contractor stalling the job because your project was underbid. It's not your job to provide the contractor with working capital. Flex your muscles by showing the contractor the contract. If the contractor underbid your project, he needs to be held accountable for his mistakes.

TRACKING EXPENSES

Writing ongoing checks in staggering amounts can be shocking, which may be why most homeowners prefer to only track expenses when there's a change to their remodeling budget. However, to understand your spending and know when a change occurs, for better or worse, you need to track every expense. Let's keep everything simple, so you can sleep better at night. Download a budget-tracking app or create a project expenses spreadsheet.

List each project expense and categorize it as either a soft cost, hard construction cost (materials and labor), or contingency, and then subcategorize it as one of the parts of a remodeling bid highlighted in chapter eleven. (More on tracking change orders in chapter 17.) For example, permits are a soft cost in the General Requirements sub-category. On your spreadsheet, include columns for the date the payment was made, to whom the check was paid, and check number. Also, every time you enter an expense it should automatically be deducted from your budgeted amount with remaining funds shown (much like a checking account) and also show the percentage of the budget that has been used for that category. For example, if your budget in the soft cost category was $70,000 and the amount spent to date in this category was $35,000, you would have spent 50% of the budgeted amount for this category. You'll also want to show the percentage of the overall budget that has been spent: if your overall remodeling budget was $550,000 and you've spent $35,000 on soft costs, you would have spent approximately 6% of your overall remodeling budget.

Project Expense	Date Paid	Payable To	Check #	Part of Remodeling Bid	Soft Cost	Construction Cost	Contingency	Total Budget
List each project expense here				*Refer to Chapter 11*				
Preliminary design	4/5	AB Design	4567	n/a*	5,981			
Schematic design	5/5	AB Design	4572	n/a*	5,981			
Design development	6/5	AB Design	4577	n/a*	5,981			
Structural engineering	6/30	XY Engineering	4584	General Requirements	24,640			
Construction documents	7/5	AB Design	4589	n/a*	5,981			
Permit 1	7/10	Your City	4590	General Requirements	1,109			
Permit 2	7/10	Your City	4591	General Requirements	1,441			
Permit 3	7/10	Your City	4592	General Requirements	68			
Permit 4	7/10	Your City	4593	General Requirements	139			
Tree removal	9/5	CD Landscaping	4601	Site Construction		7,702		
Change Order - Remediation of undetected mold	9/10	Mold R US	4607	Thermal and Moisture Protection			3,000	
Total					**$51,322**	**$7,702**	**$3,000**	**$62,024**
Budgeted Amount*					$159,500	$308,000	$82,500	$550,000
Amount Remaining					$108,178	$300,298	$79,500	$487,976
		% of Budget			32%	3%	4%	11%

The above entries are samples for illustration-purposes only. Your entries will vary.
* Design will be part of a remodeling bid under General Requirements if the contractor is providing design services

16. STAY UPDATED

Learn how to
avoid expensive glitches.

As your remodeling project moves forward, you and your team should schedule regular status meetings to keep everyone in the loop. Even if you're living in the house while the work is being done, don't rely on hallway conversations and driveway updates.

If the project is small, your status meeting might be fifteen minutes once a week to discuss what work will be accomplished over the next few days. Ask your contractor to point out exactly where he is on the project timeline, explain any delays, and tell you what to expect going forward.

Weather, material shortages, tardy shipments, and illness can introduce problems that require contingency planning and work-arounds. If an unavoidable delay—such as a week of rain—means the next step can't happen as planned, a status meeting gives the contractor a chance to let you know what they'll be doing instead to keep things moving. If rain means outdoor work can't be done, perhaps there is inside finish work the crew can do. If a delay in materials means the intended item can't be installed, ask if there's something else the crew can work on in the meantime.

On the following page is a suggested status meeting agenda.

One of the hallmarks of a good meeting is preparation and accountability. The person running the meeting (ideally the contractor or construction manager) should prepare for

STATUS MEETING
AGENDA

(if any, or questions not yet addressed)

Outstanding:

New:

On-Site:

PROJECT STATUS

Work completed in
the previous week

....

Work planned for the
next two weeks

....

Change orders

....

Schedule:
– Compare schedule to plan.
– Identify potential problems that may cause delays.
– Discuss options to get back on schedule.

....

Budget:
– Compare schedule to plan.
– Identify problems that may put the project over budget.
– Discuss options to reduce costs.

....

Quality:
– Identify work that does not conform to quality standards.
– Propose remedies.

....

Pending requests for
information

....

Payments:
– Lien releases
– Progress payments

FOLLOW UP

Meeting Summary

Reconvene date

ADJOURN MEETING

the meeting by gathering information and/or data from key team members and analyze it to provide an accurate progress update. It's helpful to give those team members the meeting agenda one or two days before the meeting. If any of the agenda items need follow-up, each item should be assigned an "owner" who will report on the item at the next meeting.

It's also a good idea to have someone take notes during the meeting and send a recap via email, ideally within a day or two. This way, you'll have a record of who was in attendance and the specifics discussed. Use the meeting agenda as an outline for taking notes and preparing the recap, following the order of the meeting. The meeting recap should also include any action items identified, their respective due dates, and an explanation for the rationale for any action taken so the "what" and "why" have been duly noted. This correspondence can be easily archived for future reference.

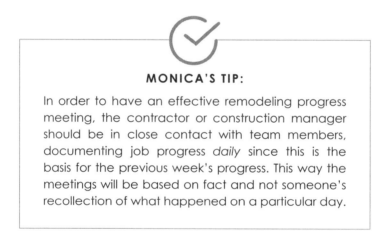

MONICA'S TIP:

In order to have an effective remodeling progress meeting, the contractor or construction manager should be in close contact with team members, documenting job progress *daily* since this is the basis for the previous week's progress. This way the meetings will be based on fact and not someone's recollection of what happened on a particular day.

Staying in the loop will pay big dividends because fewer items will fall through the cracks. Should a costly problem arise, it's always better to know sooner rather than later so you and your project team can resolve it quickly.

17. MANAGE CHANGE ORDERS

*Learn how to stay on budget
and on schedule when your plan
requires adjustments.*

Change orders happen; it's a fact of remodeling life. But thanks to your clear vision and project plan, you are much better equipped to avoid costly and unnecessary change orders. The changes you encounter are much more likely to be necessary ones, such as the need to problem-solve after finding mold behind a wall or discovering previously undetectable foundation issues. Fortunately, the process of handling change orders is not complicated.

Managing change orders effectively via formal policies and procedures helps minimize cost overruns and construction delays. The best way to do this is by documenting each change order's cost and noting the effect on the construction schedule.

REASONS FOR CHANGE ORDERS

Another reason to be thoroughly familiar your construction contract is to recognize what constitutes a change order. Following are the most common reasons for making changes midstream.

Errors. Sometimes, despite everyone's care in creating the plans, human error raises its ugly head, and fixing the problem is likely to require changes to the original plan. As soon as the problem comes to light, focus on how to fix it and what the fix will cost. Getting back on track with minimal delay and additional cost is more important than stewing over missteps. If the error was your fault, own it and move on. If the error was the fault of the contractor or a sub, discuss what, if anything, can be done to mitigate the added expense and inconvenience.

Omissions. In this case something important has been left out. No matter how many people have checked the plans, it's human nature to skip a crucial detail. When an omission is discovered, it has to be set right, or problems will multiply.

Unforeseen site conditions. No one can be completely certain of a site's true condition until the crew has started their work and gotten a good look at hidden areas. That's when unexpected problems are often found: mold, asbestos, lead paint, radon, faulty wiring or plumbing, broken structural beams, unstable fill, and more. These kinds of issues bring a project to a halt until the extent of the problem is known. There's no choice except to resolve the problem, so you can expect it to impact the budget and timeline.

Regulatory requirements. Building codes protect public health and safety and change periodically. In some states, to get a remodeling permit you are required to bring all areas of your home up to code, even if those areas are not being altered, and even if they met code at the time they were built. For example, adding a new wing of a house might not only require that the electrical box meet the demand for the additional rooms, but depending on the age of the house, might require a whole-house rewiring to meet modern code. Other areas where this can occur include smoke detectors/radon gas alarms/CO detectors, plumbing systems, septic tank systems, and furnace ductwork.

Owner-requested changes. Sometimes, despite all your planning, you realize that you just can't live with a feature that is part of the plan—or without one that isn't included. You'll need to issue a change order to track the alteration, and you and your contractor will need to discuss the impact on budget and schedule.

Materials or equipment availability. Natural disasters or political unrest somewhere in the world might mean that the wood flooring or granite countertops you ordered are suddenly unavailable. A manufacturer might discontinue a line of fixtures without warning, or go bankrupt, leaving orders unfilled. A hurricane that destroys a warehouse or washes out roads may mean shipments won't arrive on time. Many contractors rent heavy equipment, and breakdowns, bad weather, or scheduling conflicts sometimes mean borrowed equipment will be delivered late. A good contractor will explore contingency plans and tap her network for resources, but sometimes delay or substitution is inevitable, and you'll need to confer on how those changes affect your project.

Contractor/designer-requested change. An experienced contractor doesn't view a job as paint-by-numbers, even with a detailed plan. Sometimes, opportunities for options arise once a contractor gets on-site and sees the work unfolding. These might be small tweaks, like realizing that moving an outlet a few feet will lead to easier use, or it might be a bigger change based on a desire to avoid regulatory problems. Value the contractor's suggestions and then make a decision based on how the change will affect your project.

Schedule delay. Any of the previously mentioned issues can cause a schedule delay. Sometimes when a subcontractor is delayed or materials or equipment don't arrive on time, work can progress in other areas. In other cases—for example, days of rain when the roof is to be installed—a delay is unavoidable.

Don't rely only on your contractor to keep an accurate record of change orders. Get change orders, along with associated costs, in writing and approve them before the work is performed. Make sure an explanation of why the change is necessary, and its effect on your project schedule, is documented. Track change orders using Excel or a budget-tracking app, showing the link between the change and the associated costs. A timeline graphic showing the scheduling impact is also helpful. By managing change orders, you'll be able to control costs.

CHANGE ORDER LOG

Original Contract Amount $329,000.00

CO #	Reason #	Description	Time Ext. (# Days)	CO $ Amount	Cum CO $ Amount	Cum Contract $ Amount
--	--	Original Contract Amount and # Days	120	0.00	0.00	$329,000.00
1	D	Permits	0	-107.29	-107.29	328,892.71
2	C	Electrical upgrade, entire house	10	4,000.00	3,892.71	332,892.71
2	C	Plywood entire roof	0	1,500.00	5,392.71	334,392.71
2	G	Smooth stucco entire house embedded with color	0	2,500.00	7,892.71	336,892.71
2	G	Stone veneer front of house & steps	0	3,500.00	11,392.71	340,392.71
2	G	Tile steps and landng	0	500.00	11,892.71	340,892.71
2	G	Entry colums and stucco half wall	0	500.00	12,392.71	341,392.71
2	E	Two 10" Solatubes (powder and master closet)	0	932.87	13,325.58	342,325.58
					13,325.58	342,325.58
					13,325.58	342,325.58
					13,325.58	342,325.58
					13,325.58	342,325.58
					13,325.58	342,325.58
					13,325.58	342,325.58
					13,325.58	342,325.58
					13,325.58	342,325.58
					13,325.58	342,325.58
					13,325.58	342,325.58
					13,325.58	342,325.58
					13,325.58	342,325.58
					13,325.58	342,325.58
					13,325.58	342,325.58
					13,325.58	342,325.58
					13,325.58	342,325.58
					13,325.58	342,325.58
					13,325.58	342,325.58
					13,325.58	342,325.58
					13,325.58	342,325.58

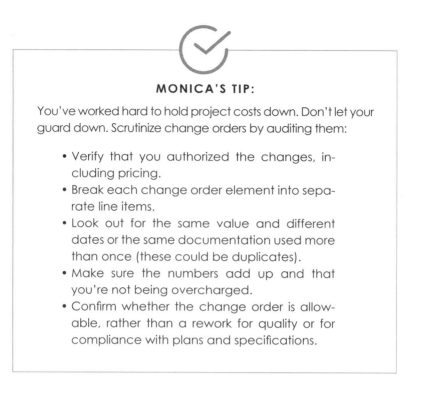

MONICA'S TIP:

You've worked hard to hold project costs down. Don't let your guard down. Scrutinize change orders by auditing them:

- Verify that you authorized the changes, including pricing.
- Break each change order element into separate line items.
- Look out for the same value and different dates or the same documentation used more than once (these could be duplicates).
- Make sure the numbers add up and that you're not being overcharged.
- Confirm whether the change order is allowable, rather than a rework for quality or for compliance with plans and specifications.

Changes can be stressful and expensive, but they occur no matter how carefully you and the contractor have planned. Don't let change orders be a source of budget anxiety. Remember, the basis for change order pricing should be part of your construction contract, and you've budgeted for contingencies. Manage the changes and track the impact, but don't let change orders sour your relationship with your contractor or dampen your enthusiasm.

18. RESOLVE DISPUTES

*Learn the five crucial steps
to take when your contractor
isn't listening to you.*

Murphy's Law states, "Anything that can go wrong will go wrong." Should a dispute arise, keep in mind that your contractor may be more familiar with solving technical problems than with settling customer service issues. But it's in everyone's best interest to solve a problem before it escalates. Often, disputes arise from a misunderstanding and become entrenched when emotions rise. If you approach the problem dispassionately, you might be able to save time, energy, money, and maybe even your relationship with your contractor. Let's explore the five options for resolving disputes.

INFORMAL DISCUSSION

This is the best place to start. Arrange a face-to-face meeting with the contractor. Have your contract, change orders, and other documentation with you, and put together detailed notes about what you need to cover. If a spouse or someone else has been very involved in the process and privy to prior discussions about the project, you may want to have them present as well, though avoid the appearance of ganging up on your contractor.

Give your contractor the chance to explain and offer solutions. Make sure the tone remains neutral and results-oriented. You may also have suggestions about how the situation can be resolved to everyone's satisfaction. Avoid seeking to blame or punish, and keep your attention on how to solve the problem.

MEDIATION

Mediation brings in a neutral party to guide you through the discussion. You may want to bring in a professional mediator, or use a trusted person with good facilitation skills. Having a mediator can help to keep tempers in check and facilitate a productive exchange, especially if prior attempts to talk things through have failed. A mediator will give each party time to tell their side of the situation, identify the key facts, try to facilitate understanding, and encourage both parties to find an amicable solution.

In selecting a mediator, choose someone both parties can trust, someone who doesn't have a personal stake in the resolution and isn't biased toward one side or the other. Mediation is not legally binding, so it can't force either party to accept a solution, but it can help you work out a plan to solve the problem. If you hire a professional, the fee will vary depending on municipality and length of the mediation.

ARBITRATION

In arbitration, a paid arbiter handles the mediation. Some contracts require arbitration in lieu of taking a case to court ("binding arbitration"), while others may require arbitration as a first step. It's wise to know what your state and municipality require before signing the construction contract, so you are aware of your options if anything goes wrong. Again, the time involved and the cost depend on your location and the length of the process. The American Arbitration Association is a good place to learn more about arbitration.

SMALL CLAIMS COURT

Small claims court generally handles cases valued at $5,000 or less, depending on your jurisdiction. As with other forms of remediation, you'll want to do your homework, be able to clearly and concisely state the problem, and have a specific solution in mind that you want the contractor to provide. The judge hearing your case won't have prior knowledge of your remodeling project, so you'll have to make the problem clear without telling a long story. You'll also need to make sure you can explain the damage incurred in a way that justifies the resolution you are seeking, and why you hold the contractor responsible. Realize that the contractor will also have a chance to tell his side of the story, so be prepared to accept some responsibility if your change orders or muddied directions contributed to the problem. Small claims court fees vary by location. Court dockets are busy, so this process could take months to resolve.

LAWSUIT

Retaining a lawyer and going to court is your most expensive option, and should be your last resort. Both you and your contractor will need to hire lawyers to represent your interests. At that point, neither of you will be able to talk to each other anymore, and the judge's decision will be binding. The loser may be required to pay the winner's court fees. Court cases can be long and drawn out, sometimes lasting for years and costing thousands of dollars.

Make sure you are certain about going down this path, and that your issue is more than hurt feelings or a quibble about details that don't really matter. That's because the facts of your case will be laid out for a judge to decide, and that judge is only going to look at what's in black and white. If you have been defrauded, if the workmanship is shoddy, if contractually promised items were not installed or delivered, you've got a case. If the real issue is that after putting everything you wanted down on paper and seeing it completed in real life, you now think you'd really rather have something else, your contractor is likely to win the day, and maybe collect monetary damages too.

No one wants to be unhappy with their remodeling project. Sometimes, plans go awry or people let you down. If that happens, it's important to understand your options and be prepared to protect your interests.

PART 3

WRAP-UP

19. CLOSE IT OUT

*Learn the four best practices for
closing out your project.*

Even though it's the last major milestone, the close-out should be anticipated from day one, because it does not happen overnight. You'll need to create a project close-out notebook, verify that the scope of work has been performed, confirm that all your expectations have been met, and close out your contract and financial commitment.

For commercial construction projects, the general contractor is typically required to provide the owner with a close-out manual that contains documents such as project history; warranties (contractor and equipment); as-built drawings (especially if the final result differs from the plans originally approved by the municipality); operating manuals for installed items such as the HVAC system; faucets, lighting fixtures, and electronics. I recommend that you create a close-out manual for your project even though it is not a requirement for residential work. This way, you'll have all of this information at your fingertips for ease of reference in the future. I also recommend future-proofing your notebook by digitally archiving the hard copies of each document.

Chapter sixteen discussed the importance of staying informed with regular project status meetings, recapped in writing. Use those logs to start compiling your close-out notebook. Here are other documents you'll want to keep in the notebook:

CLOSE-OUT CHECKLIST

Keep track of what should be in your notebook with a checklist. Once you've placed each document into the notebook, you can check the items off your list.

Key project contacts. List the name, company, role, address, phone numbers (cell and office), and email address of key team members, including subcontractors and vendors.

Contracts. Include your contract with the general contractor as well as those with other professionals you worked with, such as the designer.

Budget. Include your final budget showing tracked expenses as discussed in chapter fifteen. This gives you a snapshot of where every remodeling dollar went.

Change orders. Include the change order paperwork, along with a log that contains the reason for each change; a brief description of the change; the estimated extension to the schedule, if any; the cost of each change order, the cumulative cost of all change orders; and the total amount added to the contract bottom line. This change order one-sheet will save you from wasting time sifting through each change order in your notebook to find what you're looking for.

Payment history. Keep a log that confirms payments were made, including date, payment type, company or person paid, amount, and check number if paying by check. I also recommend keeping a copy of the cancelled checks that have cleared your account and filing the copy in this log. Print both sides of the check so you have a record of the signature of the person who cashed it.

Lien Releases. Since lien releases are issued in conjunction with payments, attach the lien release to a copy of the cancelled check or other payment verification.

Permits. Even though municipalities have requirements for record retention on their end (check to see what the requirements are in your area) it's always a good idea to retain your own records of building and demolition permits and inspections.

Plans. The set of plans originally approved by the city.

Notice of substantial completion. The contractor (or the architect) submits this form to certify that work has been substantially completed per the construction contract documents and that only minor items remain to be completed. This typically signals that it's time to schedule a final inspection so a certificate of occupancy can be issued, when applicable. The date of the document also ends the general contractor's liability for liquidated damages.

Certificate of occupancy. The municipality that issued your building permit issues this document. It's important because it indicates that your home is safe to use because all code related issues have been addressed properly. The date of this document determines the date the contractor's warranties kick in.

Equipment training. You'll want to make note that someone trained you on the use of newly installed equipment, such as a new fireplace, HVAC system, or audio-visual system.

Equipment manuals. Each piece of new equipment should have come with a manual containing the warranty, along with information regarding service, repair, and replacement parts.

Maintenance schedule. It's helpful to have a page that lists the maintenance schedule and date of the next service. This way you won't have to dig through each manual to figure it out.

List of finish products and materials. This list should include the type, color, style, grade, size, and brand of flooring, appliances, light fixtures, cabinets and cabinet handles or pulls, countertops, sinks, faucets, tubs, tile, vanities, outlets and light switches, interior and exterior doors and knobs, plumbing fixtures, alarm system (company and type), hot water heater, insulation (including rating), windows and window coverings, roofing, exterior lighting, hardscape or exterior masonry, etc. You get the point. Leave no stone unturned.

Color swatches. If you ever need to touch up the paint in a room, you'll be happy to have the corresponding paint chips on hand. If any were custom blended, include the color formula.

Material samples. Label tile, fabric, and other non-paint samples with their details so you'll have them if you need to replace them. Store them in a clear plastic envelope.

As-built drawings. This is the final set of remodeling drawings and should reflect any deviation from the original plans. The contractor should have noted any changes at the time they occurred to ensure accuracy.

PUNCH LIST

Your contractor is almost finished and there are still some unfinished tasks. What are they? How do you know? This is the time to prepare the punch list. Go over the contract and the change orders and list any items/activities that have not been finished to your satisfaction. Look for bad paint jobs, unevenly hung cabinet doors, nail pops, and any other discrepancies that make your job look less than professional. Be alert for damaged cabinets, light fixtures askew, clean-up that wasn't done properly. You paid for professional-caliber work, and your contract stipulates the completion of quality work. Add anything that needs to be fixed or touched-up to your list.

If you were waiting on any late-delivered items such as window grids or other finish-quality pieces, make note of what's missing and when it will be delivered and installed.

Has the indoor clean-up been done well (washing windows, vacuuming plaster dust, etc.)? Have dumpsters and portable toilets been removed? If not, add them to your list.

Here's what happens next and what you should have in your possession before you issue the final payment.

1. Type up your list, which should include your final payment and the mechanic's lien waivers for the general contractor and sub-contractors.

2. Schedule a walk-through with the contractor. Bring the list, a clipboard and tablet to take notes, a camera to take photos of issues that need to be resolved, and blue painter's tape to mark and number the punch list items.

3. Visit every item on your list with the contractor. Place the blue painter's tape on the item and number it based on your list so it's easy for the contractor to find and check off once completed.

4. Email a copy of your punch list to the contractor and ask her to review it and respond with a schedule of when the items will be addressed in the following two weeks (the contractor may need to bring back subs that have moved on to other projects).

5. When the punch list items are completed, set a date for a second walk-through. In some cases, several walk-throughs might be necessary.

MONICA'S TIP:

When scheduling the punch list walk-through, allow the contractor sufficient time to complete the unfinished work before you need to move back into your home.

Don't be shy. This is your project, and you should be happy with the results. Your final payment is your leverage. Don't feel pressured to sign off on poor-quality workmanship. Remember, the contractor has a stake in making sure you are a satisfied customer, since she will want your referrals and may want to take portfolio photos. Once you have made the final payment and any liens have been lifted, file the completed punch list with the contract documents.

CONTRACT CLOSE-OUT

Closing out the contract is a way of verifying that all parties have fulfilled their obligation. This is also a chance for your contractor to turn over critical documents, such as the as-built drawings. As-built drawings incorporate any changes from the original plans. They are important because they represent your project's legacy. Unless there have been drastic changes, handwritten notes and an original, marked up by the person who executed the change, is sufficient.

Any appliances or equipment installed would have come with operating manuals, as well as any warranties. Before filing these manuals, make sure you've received this information and not a sales brochure. Also, collect any signed, original written warranties from all subcontractors and suppliers as well as any test reports for soils, asbestos, energy efficiency, etc. Don't forget to collect any extra material such as paint, tile, and flooring that was not used (heck, you paid for it). You'll take pride in knowing a replacement is on hand if you need it.

FINANCIAL CLOSE-OUT

In addition to closing out your project contractually, you need to tie up loose ends financially. Confirm that all change orders have been completed and paid. Also, collect the final invoice and any outstanding lien releases, including the final one. Review all of these documents for accuracy before making final payment.

As you can see, closing out your project does not happen overnight. That's why you should start your project with close-out in mind. Better to gather the documents you need along the way showing that project deliverables have been completed and accepted, than waiting until the end to play catch-up.

20. CELEBRATE YOUR SUCCESS

*Learn how to make the most
out of your successful remodel.*

Congratulations! You've crossed the remodeling finish line and all your hard work has paid off. Those weeks or months of research, reference-checking, writing up plans, and applying for financing, going over blueprints and managing details are all behind you, and your dream project is now a reality.

Take time to savor your success. You've accomplished something that eludes many homeowners—the happy conclusion to a remodeling project realized without budget trauma or interpersonal drama. Take photos, soak up the atmosphere, and invite friends over to appreciate the outcome of your remodel. You've earned it.

Make sure your remodeling project stays in great shape, and get the most out of the finishes, materials, and appliances by taking time to go over the maintenance requirements for upkeep so you can keep everything in good condition. Ask how often surfaces like wood floors or countertops should be refinished. Understand your appliance warranties so you're prepared if something needs adjustment.

If you're happy with your contractor, architect, designer, and suppliers, consider putting a positive comment on review sites like Houzz, Porch, Yelp, or Angie's List. You may even want to offer to be a reference or allow them to use photos of your 'before' and 'after.'

Now that the project is done, relax and enjoy your new, enhanced home. Celebrate your Remodel Success!